The Cosmic-Chemical Bond
Chemistry from the Big Bang to Planet Formation

The Cosmic-Chemical Bond
Chemistry from the Big Bang to Planet Formation

DA Williams
Department of Physics and Astronomy, University College London, London, UK
Email: daw@star.ucl.ac.uk

TW Hartquist
School of Physics and Astronomy, University of Leeds, Leeds, UK
Email: twh@ast.leeds.ac.uk

RSCPublishing

ISBN: 978-1-84973-609-1

A catalogue record for this book is available from the British Library

© DA Williams and TW Hartquist, 2013

All rights reserved

Apart from fair dealing for the purposes of research for non-commercial purposes or for private study, criticism or review, as permitted under the Copyright, Designs and Patents Act 1988 and the Copyright and Related Rights Regulations 2003, this publication may not be reproduced, stored or transmitted, in any form or by any means, without the prior permission in writing of The Royal Society of Chemistry or the copyright owner, or in the case of reproduction in accordance with the terms of licences issued by the Copyright Licensing Agency in the UK, or in accordance with the terms of the licences issued by the appropriate Reproduction Rights Organization outside the UK. Enquiries concerning reproduction outside the terms stated here should be sent to The Royal Society of Chemistry at the address printed on this page.

The RSC is not responsible for individual opinions expressed in this work.

Published by The Royal Society of Chemistry,
Thomas Graham House, Science Park, Milton Road,
Cambridge CB4 0WF, UK

Registered Charity Number 207890

Visit our website at www.rsc.org/books

Printed in the United Kingdom by Henry Ling Limited, Dorchester, DT1 1HD, UK

Preface

A casual glance at the night sky on a clear moon-less night reveals to the naked eye some thousands of stars and a few planets. Modern astronomy tells us that this beautiful background to our lives on Earth is a very tiny part of the Milky Way galaxy, which itself is an almost infinitesimal component of the Universe.

But there's more in the Universe than galaxies of stars, some of which have planets. In the last half-century telescopes operating in the radio, in the millimetre and sub-millimetre wavebands, in the infrared and in the ultraviolet parts of the electromagnetic spectrum have established the existence of huge clouds of gas in the space between the stars of our galaxy and other galaxies. The total mass of gas in a galaxy may be a significant fraction of the mass of the galaxy itself. In the Milky Way, the total mass of interstellar gas is about one tenth of the mass of all the galaxy's stars.

Over the last few decades, observations, particularly in the sub-millimetre waveband, have established the remarkable fact that much of the gas in these interstellar clouds is in the form of molecules. Molecular hydrogen (H_2) is the most abundant molecule in the Universe, and carbon monoxide (CO) the second most abundant. But more complex molecules—such as simple hydrocarbons of up to a dozen atoms—are found in abundance in denser regions of these clouds. The fullerenes, C_{60} and C_{70}, have been identified in at least one source, and there is much evidence (not yet conclusive) that large molecules called polycyclic aromatic hydrocarbons (or PAHs) containing in the order of one hundred carbons exist in interstellar space. Evidently, interstellar space is chemically active.

These molecules are useful to astronomers. The radiation they emit is, in effect, a probe of the physical conditions in regions of space where the molecules are found. In fact, molecules also have important active roles to play in the interstellar medium. However, these active and passive roles do not concern us here.

In this book, our interest is simply chemical. Why are interstellar clouds mostly molecular? Do the molecules survive indefinitely, or are they also destroyed in space? What are the chemical processes that can take place under the extreme conditions in the interstellar medium? Does interstellar chemistry depend on these conditions? If so, how does the variety of molecules formed in interstellar chemistry vary from place to place in a galaxy, or from one galaxy to another?

Our discussions will include many types of chemical reaction. Some chemistry occurs in the gas phase, some in surface catalysis, and some in the chemical processing of solid-state ices. Almost always, the physical conditions in which these chemistries occur are extreme departures from those we normally experience on Earth or in the laboratory.

This molecular revolution in astronomy has therefore created a huge demand for understanding of chemistry in unfamiliar physical conditions. Chemists from many backgrounds have happily engaged with their astronomical colleagues to meet these challenges, and the subject of chemistry in space is an active and rapidly developing area of modern research: called astrochemistry. The bond between chemists and astronomers is now firmly established: this is the cosmic-chemical bond.

Contents

Chapter 1
Introduction to Astrochemistry   1

1.1 Molecules in Space   1
    1.1.1 Chemical Richness   1
    1.1.2 A Variety of Sources   2
    1.1.3 Common Interests   3
1.2 The Astronomical Background: Gas and Dust   4
    1.2.1 Interstellar Environments   5
    1.2.2 Near-Stellar Environments   13
1.3 How to Make Interstellar Chemistry Happen   14
1.4 Molecules in Astronomy   15
1.5 What is in this Book?   17
1.6 A Word about Units   19
Further Reading   21

Chapter 2
Interstellar Molecular Hydrogen   22

2.1 How do we know that H_2 is present in the Interstellar Medium?   22
    2.1.1 The Spectral Features of Molecular Hydrogen   23
    2.1.2 Emission from Molecular Hydrogen   24
    2.1.3 Invisible Molecular Hydrogen in Cold, Dark Regions   26
2.2 How is H_2 formed in the Interstellar Medium?   27

The Cosmic-Chemical Bond
DA Williams and TW Hartquist
© DA Williams and TW Hartquist 2013
Published by the Royal Society of Chemistry, www.rsc.org

2.2.1 Gas Phase Mechanisms	27
2.2.2 Reactions on the Surfaces of Interstellar Dust Grains	30
2.2.3 Wilder Speculations	33
2.3 How is Interstellar Molecular Hydrogen Destroyed?	35
2.3.1 Photodestruction in the Interstellar Radiation Field	35
2.3.2 Ionisation and Dissociation of H_2 by Cosmic Rays	39
2.3.3 Reactions of H_2 with He^+	41
2.4 The H/H_2 Balance in Interstellar Clouds of the Milky Way Galaxy	42
2.5 Why is Molecular Hydrogen Important in Interstellar Chemistry?	44
Further Reading	46

Chapter 3
Chemical Routes to Interstellar Molecules — 47

3.1 Making Interstellar Molecules	47
3.1.1 Radiative Associations of Atoms	47
3.1.2 Exchange reactions	49
3.1.3 Radiative Associations with Molecules	49
3.1.4 Ionisation by Starlight	51
3.1.5 Ionisation by Cosmic Rays	51
3.1.6 Dust Grains	53
3.2 Electromagnetic Radiation and Interstellar Chemistry	54
3.2.1 Starlight	54
3.2.2 X-Rays	57
3.3 Cosmic Rays	58
3.4 Dust Grains	63
3.4.1 Molecular Hydrogen	63
3.4.2 Dust Grains as Catalysts	64
3.4.3 Chemical Processing of Interstellar Ices	66
3.4.4 Surface Reactions in Diffuse Clouds?	67
3.5 Dynamical Heating	69
3.5.1 Shocks	70
3.5.2 Interfaces	73
3.6 Conclusion	75
Further Reading	77

Contents   ix

Chapter 4
Molecules in the Milky Way Galaxy   **78**

4.1 Diffuse Interstellar Clouds   78
   4.1.1 Gas Phase Chemistry in Diffuse Clouds   80
   4.1.2 Very Large Molecules in Diffuse Clouds?   83
4.2 Dark Clouds   87
   4.2.1 Gas Phase Molecules in Dark Clouds   88
   4.2.2 Molecules in the Solid Phase—Increasing the
       Molecular Complexity   90
4.3 Star Formation   92
   4.3.1 Chemical Complexity in Star-Forming Regions   93
   4.3.2 Routes to Chemical Complexity in Star-
       Forming Regions   95
   4.3.3 Deuterium Fractionation   99
4.4 Near-stellar Environments   100
   4.4.1 Dust Formation   101
   4.4.2 Chemistry in Cool Circumstellar Envelopes   104
   4.4.3 Chemistry in Planetary Nebulae   106
   4.4.4 Chemistry in the Ejecta of Novae   107
   4.4.5 Chemistry in the Ejecta of Supernovae   112
4.5 Conclusions   116
Further Reading   117

Chapter 5
The Path to Planets   **118**

5.1 Angular Momentum and Ionisation   118
   5.1.1 The Fractional Ionisation at the Onset of Disk
       Formation   120
   5.1.2 Inference of the Fractional Ionisation and
       Deuterium Fractionation   122
   5.1.3 'Small' Grains, Anions, and Ionisation   123
   5.1.4 Ionisation at Higher Densities and in Disks   124
5.2 Very Basic Neutral Chemistry During Collapse and
   Disk Formation   126
   5.2.1 Carbon   126
   5.2.2 Oxygen   127
   5.2.3 Nitrogen   128
   5.2.4 Abundances Changed by Subsequent
       Disk Chemistry   128

5.2.5 'Complex' Organics? 129
5.3 Chemistry in More Massive Disks 129
   5.3.1 Properties of the Disk 130
   5.3.2 Species Affected Primarily by Desorption and Adsorption Only 130
   5.3.3 Species Also Affected by Gas-Phase Reactions 131
5.4 Chemistry in More Evolved Disks 132
   5.4.1 Typical Disk Properties 133
   5.4.2 Molecular Distributions 134
5.5 Comets 136
   5.5.1 Measuring the Inventory of Cometary Volatiles 136
   5.5.2 Compositional Classification Based on Optical Data 137
   5.5.3 Compositional Classification Based on Infrared Data 138
   5.5.4 Comparison with Interstellar Chemistry 139
5.6 Meteoroids 139
   5.6.1 Stardust 140
   5.6.2 Amino Acids 141
5.7 Exoplanets 142
   5.7.1 Transmission Spectroscopy of Hot-Jupiters 143
   5.7.2 Possible Super-Earth Atmospheric Compositions 143
   5.7.3 Potential Bio-markers 144
Further Reading 145

Chapter 6
A Universe of Galaxies   146

6.1 Galaxies outside the Milky Way 146
   6.1.1 Types of Galaxies 146
   6.1.2 Applying What We Know 149
6.2 Sensitivity of Interstellar Chemistry to Parameter Variations 151
   6.2.1 Galaxies with Intense Ultraviolet Radiation Fields 152
   6.2.2 Galaxies with Intense Fluxes of Cosmic Rays 154
   6.2.3 Variations in the Relative Elemental Abundances 158
6.3 Chemistry in Different Galaxy Types 161
   6.3.1 Galaxies Dominated by Massive Star Formation 161

6.3.2 Galaxies Dominated by PDRs and XDRs	163
6.3.3 Dwarf Galaxies	164
6.4 Cluster of Galaxies	166
6.5 Conclusions	169
Further Reading	169

Chapter 7
The Early Universe — 171

7.1 Cosmic Evolution Before and During Recombination	171
7.1.1 The Expansion of the Universe	171
7.1.2 Primordial Nucleosynthesis	173
7.1.3 Recombination	175
7.1.4 The Microwave Background	176
7.2 After the Recombination Era	178
7.2.1 H_2 and HD—First Stars and Galaxies	178
7.2.2 Reionisation, H_2 and Star Formation	182
Further Reading	184

Chapter 8
Why Chemistry is Important for Astronomy — 185

8.1 Molecules as Tracers	185
8.1.1 Interstellar Diffuse Clouds	185
8.1.2 Giant Molecular Clouds	188
8.1.3 Interstellar Ice	189
8.2 Molecules as Probes	191
8.2.1 Excitation of Molecular Lines	191
8.2.2 Computational Models of Interstellar Chemistry	192
8.3 Molecules as Chemical Clocks	196
8.4 Molecules as Controls	198
8.5 Conclusions	199
Further Reading	199

Chapter 9
Why Astronomy is Important for Chemistry — 201

9.1 Astronomy as a Stimulus to Chemistry	201
9.2 Gas Phase Reactions in Astronomy	202
9.2.1 Ion–Molecule Reactions: the SIFT Technique	203

9.2.2 Neutral Exchange Reactions: the CRÉSU Technique	205
9.3 Surface Reactions in Astronomy	207
9.3.1 Molecular Hydrogen Formation	207
9.3.2 Surface Reactions Forming Species other than H_2	210
9.4 Chemical Processing of Interstellar Ice	211
9.5 Conclusion	213
Further Reading	213

Subject Index   **214**

CHAPTER 1
Introduction to Astrochemistry

1.1 MOLECULES IN SPACE

We live in a molecular Universe. Chemistry is, of course, rampant on Earth, and we have known for a long time that the atmospheres of the planets of our Solar System are almost entirely molecular. But one of the most remarkable discoveries of astronomy in the last half century—in an era of astronomical discovery that is truly astounding—has been the fact that molecules exist in abundance in huge clouds of gas between the stars in our own galaxy, the Milky Way, and in the interstellar space of galaxies beyond the Milky Way. In fact, we didn't even know of the existence of such 'giant molecular clouds' until they were detected by the emission from the carbon monoxide (CO) molecules that they contain. The telescopes that made these detections were similar to radio telescopes but operate at shorter wavelengths, detecting emission in the rotational spectrum of CO at a wavelength of 2.6 mm. These clouds themselves fully deserve the adjective 'giant'; they may contain up to one million times the mass of the Sun in a single cloud.

1.1.1 Chemical Richness

By similar techniques, these clouds were found to be chemically rich; that is, they contain a variety of simple species in addition to CO, such as hydrogen cyanide (HCN), carbon monosulfide (CS), ethynyl (C_2H), cyclopropenylidene (C_3H_2), and even ions—*i.e.*

molecules with an electron missing—such as the formyl ion (HCO^+). Denser regions embedded within these giant clouds are even richer in their chemistry. For example, very small and dense regions close to young massive stars were found to contain molecular species that are larger and more complex than those found in the extended regions of the giant clouds. These denser and more localised regions contain molecular species such as methanol (CH_3OH), ethanol (CH_3CH_2OH), propanal (CH_3CH_2CHO), and dimethyl ether (CH_3OCH_3).

Some carbon chain molecules such as cyanodecapentayne ($HC_{10}CN$) were detected first in interstellar and circumstellar space, and later confirmed in the laboratory. Macromolecules such as the cage molecules Buckminster fullerene (C_{60}) and 70-fullerenes (C_{70}) have also been detected in the gaseous envelopes of old stars. There is some evidence that interstellar space includes species known as polycyclic aromatic hydrocarbons (PAHs); these may contain up to a hundred atoms with carbons in hexagonal (*i.e.* graphitic) arrays terminated by peripheral hydrogen atoms and other structures.

The chemical variety does not end there. Isotopologues of many species have been detected, mostly where hydrogen atoms have been substituted by deuterium. For example, in some places where water (H_2O) has been detected, then the substituted versions HDO and D_2O have also been detected. Similarly, all the possible D-substituted varieties of ammonia have been detected, from NH_3 through NH_2D and NHD_2 to ND_3. These hydrogen isotopologues are often surprisingly abundant, given that the cosmic deuterium abundance relative to hydrogen is in the order of one part in one hundred thousand. Other isotopologues involve carbon and oxygen isotopes. For example, all six versions of CO involving ^{12}C and ^{13}C and ^{16}O, ^{17}O, and ^{18}O have been detected.

1.1.2 A Variety of Sources

Having noted that the giant molecular clouds—and the star-forming regions within them—were chemically rich, astronomers also turned their attention to other objects in the Milky Way galaxy and detected molecular emissions in many other regions. Evolved cool stars have extended envelopes that drift out into interstellar space. These envelopes show delightfully precise patterns of chemistry, with a system of 'parent' molecules close to the star giving rise to 'daughter' species in the envelope. These cool stars and envelopes evolve after ten thousand years or so to become the beautiful planetary nebulae, which display their own unique chemistries. Stars near the ends of their lives may explode in novae or supernovae, and the *ejecta* even in these apparently hostile environments display molecular emissions of various species. *Evidently, if chemistry can possibly occur in astronomy, it will!* Perhaps the most extreme conditions in which molecules are found are in a stellar atmosphere. Although the atmospheres of massive stars are simply too hot for molecules to exist, even common stars like our Sun—whose atmosphere has a temperature of about 5800 K—show molecular spectra of CO and of H_2 in sunspots, where the temperature is slightly lower at about 4000 K.

The discovery of molecules in external galaxies beyond the Milky Way is also an exciting story, and one that is still being explored. Although we cannot resolve the structure of very distant galaxies, and the great distances mean that the emissions are very weak, observations suggest that the wealth of chemistry that we observe in different regions of the Milky Way will be repeated in similar galaxies throughout the Universe. Even more remarkably it is now possible, in certain circumstances, to detect molecular emissions from objects that are so far away that the radiation detected was emitted when the Universe was merely a few percent of its present age of 13.7 billion years. Evidently, chemistry was occurring very early indeed in the history of the Universe.

1.1.3 Common Interests

From the point of view of chemistry, the discovery of these molecules is fascinating, and the demands of astronomy have challenged existing chemical knowledge. Chemists have met that challenge superbly and have been stimulated into enormous efforts in both laboratory and theoretical work. From the point of view of astronomy, the existence of molecules in interstellar and circumstellar gas has opened a new way to study astronomical regions that were previously inaccessible to observations. This new way of studying astronomy has enabled astronomers to find out more about interstellar and circumstellar matter, the formation of stars and galaxies, and the interaction of those stars and galaxies with their environments. From molecular spectra, astronomers can deduce densities, temperatures, elemental abundances, ionisation rates and many other important parameters.

A new subject, astrochemistry, involving chemists and astronomers has developed to exploit and enrich the overlap in interests between these two areas. The Royal Society of Chemistry and the Royal Astronomical Society have combined to encourage this collaboration in the now well-established UK Astrophysical Chemistry Group. However, the subject may have even wider ramifications. Nearly all of the identified interstellar molecules are familiar in organic chemistry. Although very simple in biological terms, some of them can be recognised as the building blocks of larger molecules of relevance to biological chemistry. For example, the amino acid glycine (NH_2CH_2COOH) has been detected in the comet Wild 2 by the NASA spacecraft Starburst, and a related molecule aminoacetonitrile (NH_2CH_2CN) has been detected by conventional radio astronomy in the centre of the Milky Way. These detections have given some support to the idea that life is widespread throughout the Galaxy, and that interstellar organic molecules may provide the feedstock for a complicated pre-biological chemistry when planets form during the process of star formation inside an interstellar cloud. As yet, these ideas remain fascinating speculations.

This book, however, is focused strictly on chemical matters. The main question we want to address is this: what are the processes by which these molecules are formed and destroyed in the various astronomical locations in which they are found? To answer it, we shall need to have some specific details of the nature of those locations. The remainder of this chapter is therefore devoted to a rather general discussion on the relevant astronomical background. More detailed descriptions are given in subsequent chapters.

1.2 THE ASTRONOMICAL BACKGROUND: GAS AND DUST

Most molecular-rich astronomical regions can be described either as *interstellar* or *near-stellar*. These regions contain solids in the form of dust particles, as well as gas.

1.2.1 Interstellar Environments

The first point to establish is the elemental composition of the interstellar gas. It is almost entirely hydrogen. Only about one tenth of one percent (by number of atoms, relative to hydrogen) is in the important elements oxygen, carbon and nitrogen, taken together. Other atoms are even less abundant. The relative abundances of the elements can be measured in the Sun and in other stars and ionised regions of space. There is some variation between these measurements; the values for the Sun are often used as a standard, though of course these numbers may not apply everywhere in the Milky Way and almost certainly do not apply in other galaxies. The solar values for the relative abundances of some important elements are shown in Table **1.1**.

We see from this information that for every 10 000 H-atoms in the interstellar medium, there will be roughly 6 O-atoms, 3 C-atoms, and 1 N-atom. So the atoms that are needed to make the molecules that have been detected (and to make terrestrial planets—and us!) are really a very minor component of the interstellar medium.

In fact, from the point of view of chemistry, the situation is even more difficult, in that not all of these atoms in the minor component are actually available to make molecules. Some of them are locked (almost permanently) in interstellar dust. Observations of the Milky Way galaxy show that the interstellar gas is everywhere mixed with dust. The dust is detected because it absorbs and scatters starlight, causing a partial or total extinction of the light of distant stars, and a reddening—much as the dusty atmosphere of Earth scatters blue light more than red and makes

Table 1.1 Solar abundances by number of atoms, relative to hydrogen, of some chemically important species. Relative abundances in the interstellar medium may be rather smaller than these values.

H	1 000 000	Mg	44
He	100 000	Si	36
O	540	Fe	35
C	300	S	15
N	74	Na	2

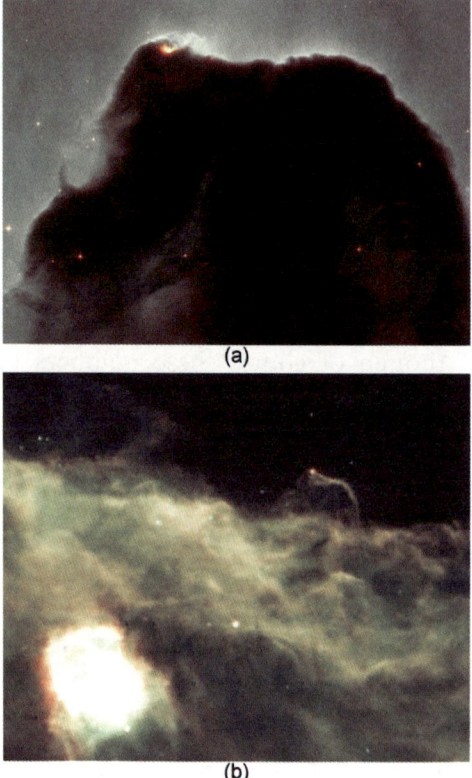

Figure 1.1 The famous Horsehead Nebula, imaged both in optical and infrared radiation. In optical light (a) Credit: NASA, NOAO, ESA and the Hubble Heritage Team (STScI/AURA); Acknowledgments K. Noll (Hubble Heritage PI/STScI), C. Luginbuhl (USNO), F. Hamilton (Hubble Heritage/STScI), the Nebula is dark and tends to obscure the light of background stars. At infrared wavelengths (b) (Credit: ESA/ISO, ISOCAM Team and L Nordh [Stockholm Observatory]) the Nebula appears bright with radiation emitted from dust grains within it.

the setting Sun appear red. The energy absorbed by dust grains also heats them, and they cool by radiating in the infrared part of the spectrum; astronomers using appropriate telescopes can also detect this radiation. Figures **1.1** and **1.2** show regions of space imaged both in visible light and in infrared radiation.

The actual nature of the dust is indicated by samples of interstellar grains identified within interplanetary dust particles collected from high-flying aircraft and from spacecraft (see Figure **1.3**). The size and composition of the grains are also

Introduction to Astrochemistry 7

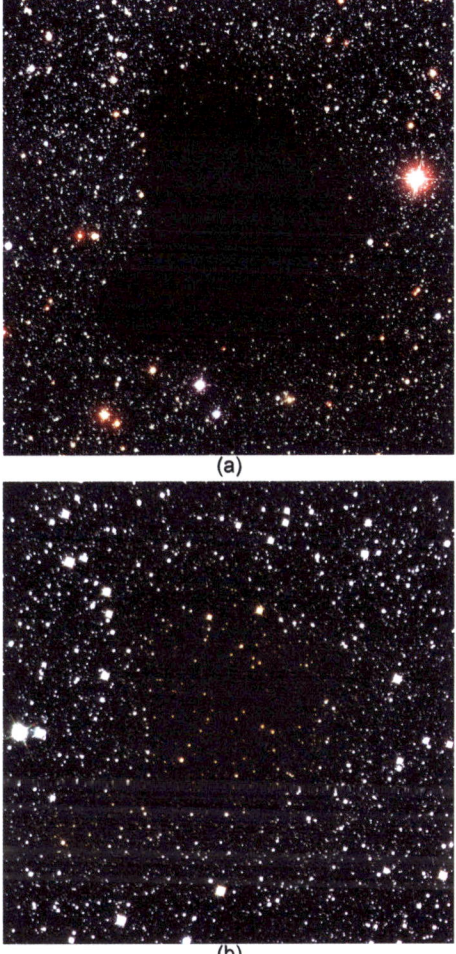

Figure 1.2 Barnard 68 is an isolated small dark cloud. Imaged at optical wavelengths (a) (Credit:FORS Team, 8.2 meter VLT Antu, ESO), it appears dark and obscures the light of background stars. At infrared wavelengths (b) (Credit: ESO), the dust grains cause less extinction than in the optical, and background stars can clearly be seen through the cloud.

inferred from models of the optics of small particles. These models suggest that the dust grains range in diameter from about one nanometre to about one micron, and the size distribution is such that there are very many more small grains than large. Their composition is basically of two main types: magnesium/iron silicates and carbons; both materials are mostly amorphous, *i.e.*

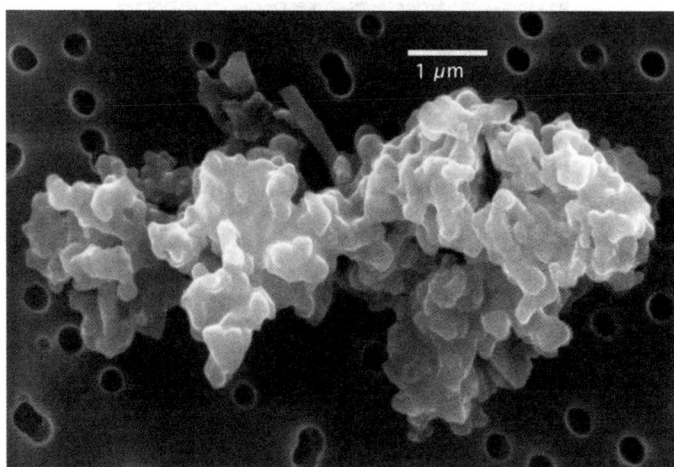

Figure 1.3 An image of a collected interplanetary dust particle (Credit: D E Brownlee). Note the open porous structure. The particle is composed of sub-micron units. The material composition is mainly of a silicate called a chondrite.

they don't have long-range order in their structures. It appears that almost all the available silicon and much of the iron is required for the dust, and about half of the available carbon. About ten percent of available oxygen is locked into silicates. So, for example, where molecules containing silicon (such as silicon monoxide SiO) are detected and found to be abundant in the interstellar medium (as they are in some locations), it may mean that grains have somehow been destroyed and the silicate material returned to the gas phase.

As we will see in more detail later, interstellar dust is crucial to interstellar chemistry. Firstly, the extinction caused by dust protects the interior regions of clouds from starlight capable of destroying molecules. Secondly, the dust grains promote reactions on their surfaces; this is particularly important for the formation of hydrogen molecules. Thirdly, in certain regions the grains accumulate molecular ices of simple species on their surfaces; more complex species are formed in solid-state chemistry. Finally, grains tend to mop up electrons, and this may influence interstellar chemistry by maintaining larger populations of ions in the gas phase. Of course, interstellar dust is crucial in other ways, too. It is the raw material from which terrestrial planets are made during star formation. All forms of life on Earth contain some atoms that were in interstellar dust.

But what of the interstellar gas itself? There is a great deal of mass in the interstellar matter of the Milky Way, over a billion solar masses of material. It is distributed in a very irregular way, probably similar to the distribution that we can see in a nearby galaxy, M51 (see Figure **1.4**).

For the Milky Way, if we imagine the total mass of the interstellar medium to be spread uniformly over the entire volume of the disk of the Galaxy, then the average number density is equivalent to about one hydrogen atom per cubic centimetre. This represents an ultra-high vacuum barely achievable in the laboratory, about thirty billion billion times less dense than the Earth's atmosphere that we are breathing, in terms of numbers of atoms or molecules per unit volume. (Of course, the Earth's atmosphere is mostly molecular nitrogen and molecular oxygen, while the interstellar medium is mostly hydrogen).

However, the interstellar gas isn't uniformly spread over the galactic disk. It is very 'clumpy', as we can see is the case in the example of the neighbouring galaxy shown in Figure **1.4**. The clouds in which the mass is distributed are almost entirely neutral and have number densities ranging from a few H-atoms per cm^3 to about a thousand in giant molecular clouds. Within those extensive giant clouds are found clumps of gas with much higher number density, perhaps ten or a hundred times denser. These clumps may be so massive that they may become gravitationally unstable and collapse under their own weight to form new stars. Finally, in the vicinity of very young stars we find even denser material with the equivalent of perhaps ten million H-atoms per cm^3. This whole range of mainly neutral density structures is embedded within very hot and very tenuous fully-ionised gas maintained by supernova explosions; no chemistry occurs in the very hot gas. This wide range of structures in the interstellar medium is illustrated schematically in Figure **1.5**.

Each of these regimes has its own physics, determined by the available source of energy. Where the gas is very diffuse, there is not much dust associated with it so starlight from massive stars can penetrate these regions. In such a situation the intense ultraviolet

Figure 1.4 The partial image is of a nearby galaxy, M51, photographed by the Hubble Space Telescope. The spiral structure is evident in the emission from bright stars. Superimposed on the image are contours of emission from carbon monoxide molecules at a wavelength of 2.6 millimetres. The carbon monoxide emission appears to trace dust lanes in M51. Reproduced with permission from S. Aalto *et al.* 1999, Astrophysical Journal **522** 165. Copyright American Astronomical Society.

light from these stars is a powerful heating and ionising source. For example, a carbon atom subjected to the typical ultraviolet interstellar radiation field would be ionised in about one century.

In denser regions there is more dust as well as gas, and therefore at positions deep within a dense cloud starlight may be almost totally excluded. In these situations the main source of heat and

Introduction to Astrochemistry

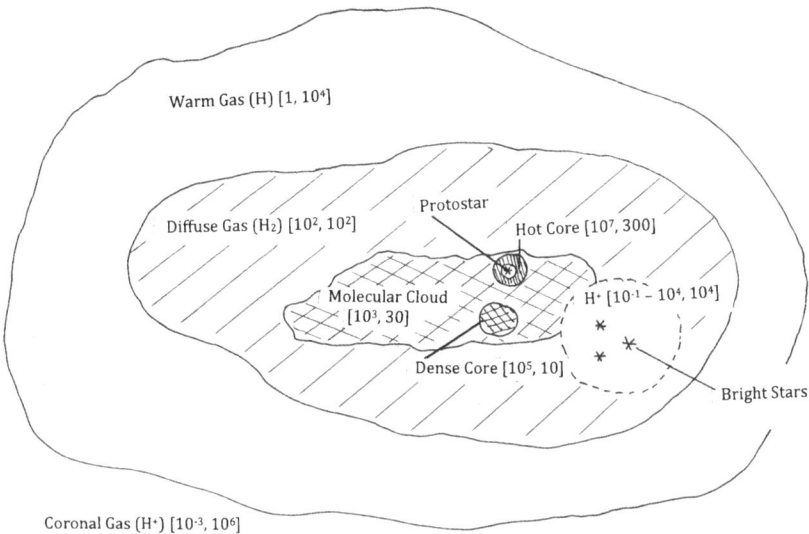

Figure 1.5 A schematic representation of different regions in the interstellar medium. Each region is characterised by two numbers: the first is the value of a typical number density of hydrogen nuclei (in whatever form) per cm^3 in that region, while the second number is the temperature of that region in K.

ionisation is the flux of cosmic rays. These cosmic rays are energetic particles, mainly of hydrogen and helium ions, that travel throughout the whole of the interstellar medium at speeds of a few tenths of the speed of light. Faster cosmic rays exist but do not contribute much to ionisation. Where starlight is excluded by dust, cosmic rays become the most important energy source. Cosmic rays are so energetic that they can collisionally ionise any atom and they have so much energy that they are essentially unaffected by the interaction. However, their flux is very low so that, even in a dense core, a hydrogen molecule is only ionised on average once every billion years.

So each interstellar region has its own physical conditions; the basic parameters of density, temperature, and extinction of starlight by dust, together with the radiation field and the cosmic rays, determine the chemistry that can occur and the molecules that can be formed. We summarise in Table **1.2** the various types of interstellar region that are important for interstellar chemistry,

Table 1.2 Some regions of chemical interest in the interstellar medium, with very approximate values of some of their physical parameters. The main heating sources are the interstellar ultraviolet radiation field (**UV**), cosmic rays (**CR**), or infrared (**IR**) from hot dust. The main chemical drivers (see Chapter **3**) are UV, CR, dust grain chemistry (**GR**), and local gas dynamics (**DYN**).

Interstellar region	Number density (cm^{-3})	Temperature (K)	Visual optical depth	Typical size (pc)	Characteristic molecule	Main heating source	Main chemical drivers
Diffuse cloud	10^2	10^2	1	10	H_2	UV	UV, CR
Dark cloud	10^3	10	5	1	CO	CR	CR
Dense pre-stellar core	10^5	10	10	0.1	CS	CR	CR, GR
Hot core	10^7	200	1000	0.01	CH_3CN	IR	CR, GR, DYN

Introduction to Astrochemistry   13

and give their typical densities and temperatures, and important molecular tracers.

1.2.2 Near-Stellar Environments

The surface temperature of a star depends very strongly on its mass. A massive star may have a surface temperature of several tens of thousands K (clearly very hostile to molecules), while the Sun has a surface temperature near 6000 K (apart from sunspots, as we have seen in Section **1.1**). Evidently, the Sun is just too hot to have molecules in high abundance, but many stars exist with masses rather lower than the Sun's mass and they have lower surface temperatures. Stars with surface temperatures in the range of about one or two thousand K are found to be almost entirely molecular in their surface layers. The number densities in these stellar atmospheres are very much higher than what is found in interstellar clouds, perhaps one thousand billion times the mean density of interstellar space. In fact, these densities are so high and the temperatures so benign that collisions between molecules enable the atoms to exchange freely in three-body collisions, so that after many collisions they combine to form the most stable molecules possible, consistent with the physical conditions. In oxygen-rich cool stars, the atmospheres contain molecules such as CO, H_2O, CO_2, and NO, while in carbon-rich cool stars they include species such as CO, N_2, C_2H_2, NH_3, and HCN.

As we mentioned in Section **1.1**, towards the end of their lives these low-mass stars begin to lose their atmospheres. The material drifts out into space, and the conditions in the gas of high density and warm temperatures may become appropriate for dust formation. Some stars are seen to vary in intensity because of the formation of dust and its dissipation as the gas in which it is embedded becomes more dilute as it flows away from the star. As it does so, the interstellar radiation field begins to penetrate the outflowing gas, and creates a new chemistry from the parent molecules, *i.e.* the molecules that were originally formed in the atmosphere, giving rise to daughter species such as HCN and more complicated cyanides known as cyanopolyacetylenes (HC_3N, HC_5N, HC_7N, *etc.*). We will return to this special chemistry in Section **4.4.2**.

As the dusty envelope drifts away, the central star may in some cases contract and become very hot (becoming a white dwarf star) so that the gas in the envelope is now irradiated from below by an intense and powerful source. This creates a new scenario for chemistry: the planetary nebula.

If a star with an envelope is in a binary with a white dwarf, some envelope material may be transferred and compressed at the surface of the white dwarf to such an extent that nuclear reactions begin and a huge explosion occurs. Astronomers detect this kind of event as a nova, an abrupt and huge brightening of a star. In the case of a single massive star, the eventual depletion of fuel in the stellar interior causes a stellar implosion followed by a 'bounce'. The resulting explosion drives *ejecta* into space at high velocity and the star becomes intensely bright, in fact, comparable in brightness to an entire galaxy of many billions of stars. It is remarkable that molecules are detected in these *ejecta*: they seem to be—and are—quite hostile environments for chemistry to occur.

Near-stellar environments have been shown to be impressive laboratories for molecule formation and for the growth of dust grains. However, as we shall see, the physical conditions and the chemical processes involved are sometimes quite different from those that occur in interstellar clouds.

1.3 HOW TO MAKE INTERSTELLAR CHEMISTRY HAPPEN

The wealth of chemistry that is found in interstellar and near-stellar environments is impressive. In fact, a very crude rule of thumb suggests that if the hydrogen density is greater than about one H-atom per cm^3 and if the temperature of the gas is less than a few thousand K, then molecules may be formed in detectable amounts. Where these conditions are comfortably exceeded (*i.e.* much higher density, and possibly a much lower temperature) then the gas may be almost entirely molecular. Of course, this depends on the local radiation field too, since radiation can destroy molecules. But, generally, as the gas density increases, so does the extinction of starlight caused by dust, so that the damage to molecules caused by radiation is mitigated.

In most of the regions of the interstellar medium that are listed in Table **1.2**, the typical temperatures are low. In these regions, the observations suggest that hydrogen is almost entirely molecular (we will return to this molecule in the next chapter). So the problem of interstellar chemistry—understanding how the molecules are made—reduces to understanding how molecular hydrogen in reaction with atomic oxygen and atomic carbon and atomic nitrogen (*etc.*) can initiate a network that produces the observed wealth of chemistry.

This presents an immediate problem: gas-phase reactions of molecular hydrogen with these atoms are strongly suppressed at temperatures of 10–100 K. There are barriers to reaction of many times the kinetic energy in the colliding partners. Of course, if the temperature is much higher say comparable to those in cool stellar envelopes, this restriction does not apply and exchange reactions between, say O-atoms and H_2 are efficient.

So, in some way chemistry in space needs to be driven. We shall return to this in Chapter **3**, but it's useful to list here the ways that nature finds to overcome the challenge of chemistry in space. Firstly, the energy in the interstellar radiation field can help by creating ions that do react with H_2. Secondly, where the interstellar radiation field is excluded by the extinction caused by dust, cosmic rays still penetrate and create ionisation that promotes chemistry. Thirdly, warming the gas can promote chemistry by allowing reaction barriers to be overcome. Finally, dust grains can promote chemistry on their surfaces and also by accumulating molecular ices of simple species that can be converted to more complex forms. In the following chapters, we will see examples of all these schemes in operation.

1.4 MOLECULES IN ASTRONOMY

Interstellar and circumstellar molecules absorb and emit radiation, and the radiation we receive from them tells us about the physical conditions of the gas—especially the density and temperature—where the molecules are formed and destroyed by the chemical processes we shall discuss in Chapter **3**. Therefore, molecules are excellent tracers of physical conditions, if we know how to interpret the information contained in the radiation. Much of the information we receive in this way is from cold and relatively dense gas; because of the dust mixed in the gas, these regions are often optically opaque, and the information that we receive comes in radiation with wavelengths in the millimetre and submillimetre; *i.e.* very short-wavelength radio waves. Molecules undergoing transitions from one rotational level to another emit much of this radiation. For example, the CO molecule, may be excited from its ground rotational level, $J = 0$, by a collision with a hydrogen molecule to the first excited level of CO, *i.e.* $J = 1$ (where J is a quantum number that measures the total angular momentum of the hydrogen molecule in multiples of $h/2\pi$ and h is Planck's constant):

$$CO(J=0) + H_2 \rightarrow CO(J=1) + H_2$$

In this collision, the H_2 molecule has given up the small amount of energy to excite the upward transition in the CO molecule. After an interval of about half a year (on average), the excited CO molecule radiates spontaneously:

$$CO(J=1) \rightarrow CO(J=0) + \text{radiation}$$

unless another collision with an H_2 molecule occurs first and de-excites the $CO(J = 1)$:

$$CO(J=1) + H_2 \rightarrow CO(J=0) + H_2$$

and carries off the energy of rotational excitation that was in the $CO(J = 1)$ molecule. Alternatively, a collision of CO with H_2 may raise the CO molecule from the $J = 1$ level to the $J = 2$ level:

$$CO(J=1) + H_2 \rightarrow CO(J=2) + H_2$$

where the energy required comes from the energy of motion of the H_2 molecule. The radiation emitted spontaneously by the $CO(J = 1)$ molecules can be detected: it has a wavelength of 2.6 mm. If detected, we can say immediately that some of the H_2 molecules must have enough energy, *i.e.* be hot enough, to excite the transition in the CO molecule, and that the number density of molecular hydrogen must not be so great that collisional de-excitation quenches the radiation or raises the CO molecule to a higher rotational level. So, without even enquiring very deeply, we already have some information about the gas where the H_2 and CO molecules exist. With the detection of other lines from CO and from other molecules, we can constrain these crude estimates much more precisely. So this kind of molecular line radiation is an excellent tracer of physical conditions in the gas. We will return to this topic in Chapter **8**.

There is also another crucial role that interstellar molecules perform. In the example described above, the radiation emitted by the CO molecule originally came from the motion of the H_2 molecule. If this radiation escapes from the gas, then it represents a loss of energy to the gas, in other words, it is a cooling mechanism. There are many situations in astronomy where external forces do work on a gas. It might be that gravity tries to compress a gas, or that a collision between gas clouds in motion compresses the gas; in situations like these a gas will tend to become hot and will stay hot unless there is some way of cooling. A hot gas has a high pressure and resists further compression. But if the gas can radiate then it can cool, so a cool gas compressed by gravity can remain cool while the density rises to much higher values. This is basically how a galaxy or a star can form. We will describe more about regions of star formation in Section **4.3**.

An inhomogeneous magnetic field also produces force that affects the dynamics of astrophysical gas. The chemistry in molecular regions controls the ionisation, which influences the evolution of the magnetic field.

The molecules are, therefore, essential components of the Universe. They allow us to trace the gas in which they are formed and to determine the physical conditions in that gas. They also control the evolution of that gas, by determining the cooling of gas subject to external forces and affecting the magnetic field and the role it plays in dynamics.

1.5 WHAT IS IN THIS BOOK?

Molecular hydrogen is the most abundant molecule in the Universe. It is also the seminal molecule for very nearly all interstellar and circumstellar chemistry. Without abundant H_2, chemistry could not supply the variety and wealth of molecules that we observe in the Universe. Molecular hydrogen can be formed in some gas phase reactions, but usually not in the abundance required. It was postulated over 50 years ago that reactions at the surfaces of dust grains might be a more efficient method of making H_2. Laboratory experiments and detailed theory have since supported that view (see Section **9.3**). The story of interstellar molecular hydrogen, its spectroscopy and detection, its formation and destruction, its role in chemistry and its general astronomical significance embraces all the important issues of interstellar chemistry. It is appropriate to use a discussion of H_2 in the next chapter to introduce them in this book.

Having explored and introduced some of the main chemical paths (with reference to H_2) in Chapter **2**, we give in Chapter **3** a more detailed discussion of the main routes by which chemistry can be initiated (as first mentioned in Section **1.3**). Here, we are concerned with relatively simple species; how are they produced and destroyed in interstellar clouds?

Chapter **4** is a large chapter covering many aspects of chemistry in the Milky Way galaxy. From an astronomical perspective, molecules help to give an insight into star formation, now considered to be one of the major topics of modern astronomy. So the first and second sections of this chapter are concerned with the formation of stars of low and high masses, and the influence that newly formed stars can have on their environments. This chapter also describes in more detail the chemistry of the near-stellar material.

Chapter **5** describes the formation of planetary systems like the Solar System in which we live. Large numbers of extra-solar planets have now been detected, and it is clear that multi-planet systems containing planets similar to the Earth exist in orbit around stars other than the Sun. What roles do molecules play in the formation of planetary systems? Obviously, this is—at present—a rapidly emerging area of study. There are also implications for astrobiology, and we briefly mention them.

Chapter **6** looks outside the Milky Way to the Universe of galaxies. These galaxies may have physical conditions that depart from those in the Milky Way. They may have very high star formation rates—possibly triggered by collisions between galaxies—and some galaxies have very active nuclei where material is being swallowed by massive black holes, generating intense radiation fields. These differences from what is, by comparison, the sedate Milky Way affect the chemistry and the molecules that can be used as tracers in these active galaxies. Most galaxies are so distant from us that they are unresolved, so molecular emissions from all the various features such as those that we find in the Milky Way cannot be spatially distinguished. Is it possible to use information in molecular emissions to infer the nature of the unresolved galaxy?

Chapter **7** looks back to an era before galaxies had formed and before the formation of stars. The Universe was chemically very simple then. Did chemistry have a role to play? Molecules could have been important in allowing gas to cool and hence—possibly—to collapse under gravity.

Chapters **8** and **9** conclude the book by looking in some detail at how chemistry has benefited astronomy and at how chemistry has also gained by responding to the demands emerging from astronomy. It is not an exaggeration to say that molecular emissions have revolutionised the view that astronomers have of the Universe. It is perhaps less recognised that the needs of astronomy have led to extensive studies of gas-phase, gas-surface, and solid-state chemistries, involving both state-of-the-art experiments and fundamental quantum mechanical computations. The development of astrochemistry has been a mutually beneficial activity for astronomers, chemists, and for those who now consider themselves astrochemists.

1.6 A WORD ABOUT UNITS

In this book, we shall be mostly concerned with understanding the nature of chemical processes that occur in astronomy. We shall not be writing down mathematical equations, nor making complex calculations. Occasionally, we shall have to refer to a quantity in some appropriate units, and so we discuss briefly here the (slightly non-standard) units in use by astronomers.

There are some areas of science where the universal use of SI units is not helpful. On the large scale, there are obvious problems. For example, one can refer to the mass of the Sun as 1.989×10^{30} kg. This is an important mass, but it is more convenient to refer to it simply as one solar mass, and to measure the mass of other stars in terms of solar masses. Similarly, the distance to the Sun from the Earth is 1.496×10^{11} m, and again, this is a fundamental distance (called one Astronomical Unit) to which other astronomical distances can be compared. The distance to an object that subtends one second of arc across a diameter of the Earth's orbit about the Sun is said to be one parsec, which is in SI units 3.086×10^{16} m, or about three light years. Distances in astronomy are therefore related to the diameter of the Earth's orbit through this unit of the parsec. Distances to stars in the Milky Way galaxy are therefore measured in parsec and kiloparsecs, and distances to other galaxies in kiloparsecs and megaparsecs.

Astronomers have been rather slow to use SI units for quantities on the atomic scale. For example, number densities are generally given in terms of a number of atoms or molecules per cm^3, rather than per m^3. The only possible justification for this reluctance on the part of astronomers to switch from cm^3 to m^3 is that the mean number density of interstellar matter averaged over the volume of the disk of the Milky Way galaxy is about one H-atom per cm^3. Therefore, when one refers to a diffuse cloud with number density of 100 H-atoms per cm^3, or a dark cloud with number density of 10^4 hydrogen molecules per cm^3, or a dense core of gas near a new star with number density of 10^7 hydrogen molecules per cm^3, one is looking at a number that measures directly how dense this particular region is compared to a fundamental density for the galaxy. This is highly convenient, and so we shall generally use number densities defined in atoms or molecules per cm^3. Obviously, numbers per m^3 are simply a million times larger.

Given that number densities are stated by astronomers in this form, then rate coefficients are stated in appropriate units; for example, binary reaction rate coefficients are given in terms of $cm^3 \ s^{-1}$ and ternary reaction rate coefficients in terms of $cm^6 \ s^{-1}$.

In terms of energies and wavelengths, astronomers tend to use a mixture of units. Wavelengths in the optical and ultraviolet used to be quoted in Angstroms but are now most often given in nanometres; at longer wavelengths micrometres and millimetres are used, with centimetres and metres in the radio region. Astronomers are often concerned with atomic and molecular spectra and quote energies in terms of electron volts (eV) or equivalently in Kelvin (K), rather than an inconvenient number of Joules. Temperatures are, of course, measured on the absolute scale in units of Kelvin, with zero at -273.15 on the Celsius scale.

We show in Table 1.3 some useful conversions between the hotchpotch of units used by astronomers and more standard usages of other branches of science.

Introduction to Astrochemistry

Table 1.3 Conversion factors between some units commonly used in astronomy and conventional SI units.

Mass	1 Earth mass	5.972×10^{24} kg
	1 Solar mass	1.989×10^{30} kg
Length	1 Astronomical Unit	1.496×10^{11} metres
	1 light year	9.5×10^{15} metres
	1 parsec	3.08×10^{16} metres
Time	1 year	3.156×10^{7} seconds
Energy	1 electron-volt	1.6022×10^{-19} Joules
	1 electron-volt	1.1605×10^{4} Kelvin
	1 Kelvin	1.3806×10^{-23} Joules
Wavelength	1 nanometre	10^{-9} metre
	1 micrometre	10^{-6} metre
	1 millimetre	10^{-3} metre
	1 centimetre	10^{-2} metre

FURTHER READING

T. W. Hartquist and D. A. Williams, *The Chemically Controlled Cosmos*, Cambridge University Press, UK, 1995 (an elementary introduction to interstellar chemistry).

A. G. G. M. Tielens, *The Physics and Chemistry of the Interstellar Medium*, Cambridge University Press, UK, 2005 (a comprehensive advanced text).

B. T. Draine, *Physics of the Interstellar and Intergalactic Medium*, Princeton Series in Astrophysics, USA, 2011 (a comprehensive advanced text).

CHAPTER 2
Interstellar Molecular Hydrogen

2.1 HOW DO WE KNOW THAT H_2 IS PRESENT IN THE INTERSTELLAR MEDIUM?

Although molecular hydrogen is the most abundant molecule in interstellar space, it was not the first molecular species to be discovered in the interstellar medium. It was identified in 1970. But about three decades earlier, optical molecular absorption spectra had been discovered in the spectrum of a hot star. It was immediately clear that these absorptions did not originate in the star but in interstellar space. Shortly afterwards, these optical spectra were assigned to the methylidyne radical (CH) and its ion (CH^+), and the cyanogen radical (CN). These species exist in what we now call diffuse clouds (see Chapter **1**, Table **1.2**). At the time, these detections and assignments caused some surprise, as it had been thought that the interstellar medium was much too hostile for molecules to exist in detectable quantities. We discuss the chemistry of diffuse clouds in Section **4.1**.

The discovery of molecular hydrogen had to wait until spectrographs operating in the ultraviolet part of the spectrum could be carried above the Earth's atmosphere in rocket-borne experiments. For hydrogen molecules, like hydrogen atoms, have strong electronic transitions in the ultraviolet, and none in the optical part, of the electromagnetic spectrum. Massive stars are hot enough to emit in the ultraviolet and make excellent background continuum sources for ultraviolet spectroscopy. Any molecular hydrogen in the line of sight between the star and Earth absorbs starlight at the signature wavelengths for H_2.

2.1.1 The Spectral Features of Molecular Hydrogen

In fact, the spectrum of a molecule like H_2 is more complicated than that of an atom such as H. Atoms and molecules both have electronic transitions in which energy is absorbed or emitted when the electrons transfer to a higher or lower energy state, respectively. But the nuclei of molecules are not fixed; they move. They can vibrate to and fro, and they can rotate end over end in the case of a diatomic like H_2. Just as the electronic energies are confined to specific values so that transitions occur only between those energies, so it is with molecular vibration and rotation. Each electronic state has a 'ladder' of permitted vibrational levels attached to it, and each of those vibrational levels has another 'ladder' of permitted rotational levels attached to it. The 'steps' of the vibrational ladder are much closer than the jumps between the electronic states and the lower rotational 'steps' are (at least initially) much smaller than the vibrational 'steps'.

The electronic transitions in molecules such as H_2 occur between states with specific electronic, vibrational and rotational energy. So instead of a single line that would be obtained for an atom in transition between two electronic states, for a molecule many lines are obtained, representing transitions between the same electronic states but different vibrational and rotational energies. The observed ultraviolet absorption spectrum of interstellar H_2 along the line of sight towards a particular bright star is shown in Figure **2.1**; the vibrational and rotational structure is clearly seen.

Along the particular line of sight shown in Figure **2.1**, the fraction of hydrogen in molecular form is on the order of 0.1 percent. This spectrum was obtained along a line of sight through a diffuse cloud where the number density of molecular hydrogen is rather low and the extinction of starlight caused by interstellar dust is correspondingly small. Higher gas densities mean more dust and higher extinction of starlight. While this certainly helps to protect molecular hydrogen against photodissociation by the interstellar radiation field (see Section **2.2.1**, below), it also drastically weakens

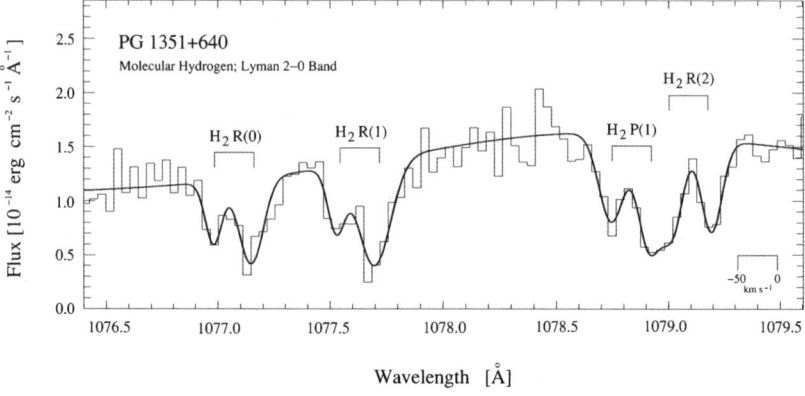

Figure 2.1 A portion of the ultraviolet absorption spectrum of H_2 in the halo of the Milky Way galaxy, the background source of radiation being a very distant quasar (PG 1351+640). The spectrum was taken using the Far Ultraviolet Spectroscopic Explorer (FUSE) orbiting observatory. There are two component H_2 clouds on the line of sight, with velocities differing by 50 km s^{-1}. The absorption lines belong to the H_2 Lyman bands corresponding to absorption from the H_2 ground electronic state in its lowest vibrational level ($v'' = 0$) to the first excited electronic state in its vibrationally excited level $v' = 2$. $R(0)$, $R(1)$, and $R(2)$ correspond to upward jumps $\Delta J = 1$ from the J level indicated, while $P(1)$ implies a downward jump $\Delta J = -1$, i.e. $J = 1 \rightarrow J = 0$ in these electronic transitions. Reproduced with permission from P. Richter et al. 2003, Astrophysical Journal **586**, 230. Copyright American Astronomical Society.

the light from the star so the absorption by H_2 cannot easily be detected.

2.1.2 Emission from Molecular Hydrogen

The ultraviolet spectrum of H_2 can also be seen in emission (as distinct from absorption) when the molecular gas is appropriately excited. Molecular hydrogen near a hot star will be excited to high electronic states by absorbing starlight. Then the excited H_2 molecules relax to the ground electronic state, emitting a variety of lines. Here, we are not looking at absorption by molecules in a particular line of sight, but emission by molecules in a volume of H_2 gas distributed around the hot star. Figure **2.2** shows the electronic emission spectrum of H_2 measured near a particular bright star as the exciting source. In such a region—close to a hot star—like that of Figure **2.2**, the radiation field is particularly

Interstellar Molecular Hydrogen 25

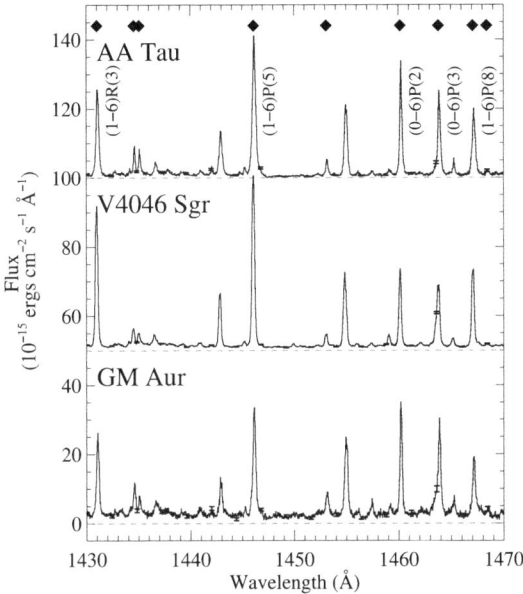

Figure 2.2 A portion of the H_2 ultraviolet emission spectrum for each of three stars, taken using the Hubble Space Telescope. The vibrational and rotational assignments of these electronic transitions are indicated (*cf.* the caption to Figure **2.1**). Reproduced with permission from K. France *et al.* 2012, Astrophysical Journal **756** 171. Copyright American Astronomical Society.

intense and controls the behaviour of the gas. Regions like this are called photon dominated regions (or PDRs).

The excitation of H_2 by stellar UV radiation (or by other means) from the ground electronic state raises the molecule to an excited electronic state; but this excited state is short-lived and the molecule quickly relaxes to the ground electronic state, emitting radiation in the ultraviolet, as illustrated in Figure **2.2**. This downward transition may leave the H_2 molecules in highly excited vibrational and rotational states of the ground electronic state. Molecules in these excited states slowly lose energy in jumps of one or two steps down the vibrational and rotational ladders, emitting radiation as they do so. This emission spectrum lies in the infrared, just beyond the red end of the optical spectrum, and can be detected in many astronomical objects where H_2 is excited by starlight; see Figure **2.3**.

After excitation by UV radiation, once the molecules have arrived in the ground electronic state and on the bottom step of the

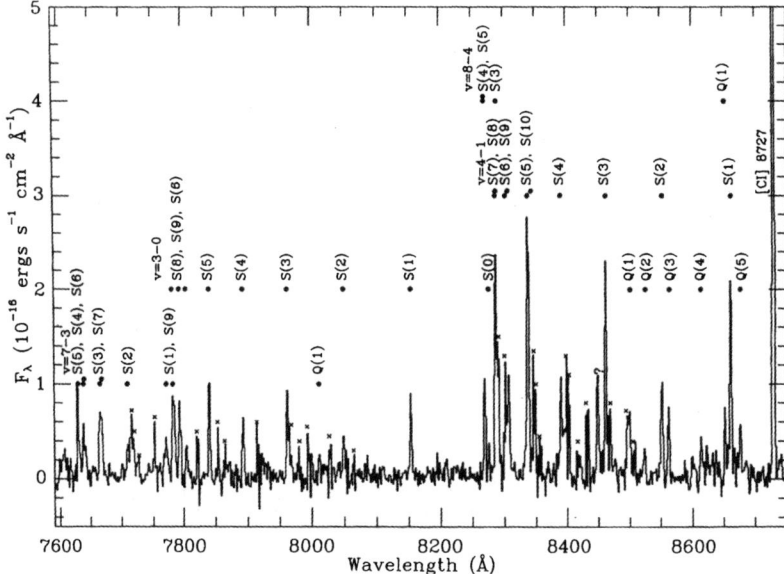

Figure 2.3 A portion of the near-infrared spectrum of H_2 in the reflection nebula NGC 2023, obtained using the Anglo-Australian Telescope. Rotational transitions within the vibrational series $3-0$, $4-1$, $7-3$, and $8-4$ are shown; $S(J)$ implies that $J \rightarrow J+2$ in the vibrational transition, and $Q(J)$ implies that J is unchanged in the vibrational transition. Reproduced with permission from M. G. Burton *et al.* 1992, Monthly Notices of the Royal Astronomical Society **257** 1P. Copyright Royal Astronomical Society.

vibrational ladder, they may still be high on the rotational ladder. A slow cascade down these rotational steps gives rise to a pure rotational spectrum of H_2, and this spectrum can also be observed in the interstellar medium. The pure rotational lines of H_2 appear further out into the infrared than the vibration–rotation lines, the longest wavelength line being at 28.2 micrometres. Collisions in a hot gas can also excite populations in vibrational and rotational levels. However, the populations and the resulting infrared spectra are different from those in the case of UV excitation.

2.1.3 Invisible Molecular Hydrogen in Cold, Dark Regions

There is plenty of evidence that H_2 is present in the interstellar medium. It is detectable in absorption in diffuse clouds, and in emission near hot stars. It is, however, not possible to detect it in dark clouds at low temperatures (see Table **1.2**), which is, in fact, where we believe most of the H_2 resides. These dense clouds are too dark to allow background stars to be used as a source for absorption spectroscopy, and too cold to excite transitions to upper levels in the H_2 from which a radiative cascade might be detected.

This failure in our ability to make direct detections of H_2 in the locations where it is most abundant is curious. We will see in Chapter **3** how astronomers overcome this problem through chemistry, by using CO as a proxy for H_2.

2.2 HOW IS H_2 FORMED IN THE INTERSTELLAR MEDIUM?

Arthur Eddington, who gained worldwide fame by leading the solar eclipse expedition that confirmed one of Einstein's predictions from general relativity, gave the 1926 Royal Society Bakerian Lecture on *Diffuse Matter in Space*. He concluded that no known mechanism could maintain non-trivial abundances of molecules in space. It was not until the 1960s that the mechanisms that form abundant amounts of molecular hydrogen in space were properly understood.

2.2.1 Gas Phase Mechanisms

(i) One might think, naively, that making molecules of hydrogen in a gas of hydrogen atoms would be quite straightforward. After all, collisions between pairs of hydrogen atoms must be common in such a gas. However, such collisions between H-atoms do not lead to H_2 molecules. The problem is that the colliding atoms still have all the energy they had as free atoms, so there is nothing to prevent them separating; in other words, they simply 'bounce' off each other.

For a colliding pair of H-atoms to form a hydrogen molecule, the pair must lose some energy. Then the individual atoms cannot separate. In principle, the pair could radiate some energy away (a reaction known as 'radiative association'). In practice, however, the colliding pair is a very poor radiator, so that the chance that the pair can radiate in the short period while the atoms are in contact is almost vanishingly small. So this 'obvious' method must be immediately ruled out.

(ii) Of course, a third H-atom might impact on the colliding pair and remove some energy, leaving the original pair bound together. This is a perfectly viable method of making H_2. It is a so-called 'three-body reaction' and is the means by which much of terrestrial chemistry occurs. But it does require a high density, so that the third body can collide with the original colliding pair before they have separated. The number density in a gas of H-atoms that is enough for three-body chemistry to begin to play a role is about 10^{11} per cm^3, *i.e.* one hundred billion H-atoms per cm^3.

This is far larger than the number density in any of the interstellar clouds we described (see Table **1.2**) although such densities might be reached in some stellar atmospheres. Clearly, molecular hydrogen is not formed in three-body reactions in a pure H-atom gas under the conditions pertaining anywhere in the interstellar medium.

(iii) There are, however, two gas phase mechanisms that mimic a three-body reaction; there are certainly three bodies involved but the reaction occurs in two stages. In the first mechanism an electron, e, attaches to an H-atom with the emission of some radiation

$$H + e \rightarrow H^- + \text{radiation}$$

This new species, H^-, the atomic hydrogen negative ion, then reacts with another H-atom, followed by the second stage

$$H^- + H \rightarrow H_2 + e$$

to form a molecule of hydrogen; in this second stage the electron carries away some energy, so stabilising the H_2 molecule. Here, the electron is effectively a catalyst that creates H_2 *via* the intermediary, H^-.

The second mechanism is similar:

$$H + p \rightarrow H_2^+ + \text{radiation}$$

followed by

$$H_2^+ + H \rightarrow H_2 + p$$

Here, a proton, p, is the catalyst that creates H_2 *via* the intermediary, H_2^+. In both cases, the intermediary is created through a 'radiative association'. In other words, the intermediary is stabilised by the emission of energy in the form of radiation. This emission is usually a relatively slow process, so the first step in each case is the rate-limiting step.

The second stage in the first scheme is called 'associative detachment' because the two atoms are associated (*i.e.* joined together in a molecule) while the electron becomes detached. The second stage in the second mechanism is an 'ion-molecule reaction' and also an 'exchange reaction', for obvious reasons. Both these second-stage reactions are fast, occurring on almost every collision.

In interstellar clouds, a small amount of hydrogen is ionised by cosmic rays. These energetic particles (mostly protons, electrons, and helium nuclei) flow though the interstellar gas. Each ionisation creates one electron and one proton. So reaction schemes involving H-atoms with electrons or with protons are certainly viable. However, these schemes are not very efficient in interstellar clouds because, firstly, the initial step in each case is rather slow, and secondly, because the intermediary in each case can be destroyed rather easily by starlight. H^- is readily destroyed by infrared light:

$$H^- + IR \rightarrow H + e$$

a process called 'photodetachment' and H_2^+ is destroyed by visible and ultraviolet light:

$$H_2^+ + Vis \rightarrow H + p$$

a process called 'photodissociation'. So, although these methods may be applicable in certain situations (especially those where there is an absence of dust, see Section **2.2.2**) they are found to be inadequate in the general interstellar medium. However, they do have an important role to play in the chemistry of the early Universe (see Chapter **7**).

(iv) There are methods that invoke 'chemical exchange' reactions of the type:

$$XH + H \rightarrow X + H_2$$

These reactions can be perfectly viable, often at very low temperatures. However, as a method of forming H_2 they merely push the problem back one stage: how does one make XH? While hydride molecules are observed in the interstellar medium, they do have a low abundance relative to hydrogen, typically one part in 10^8. This means that the formation rate of H_2 by this route can never be large, and certainly never large enough to account for the amount of interstellar H_2 that is observed.

Although this discussion of gas phase formation of H_2 has not been successful in identifying an efficient method, all of the reaction types introduced here appear much more positively in the wider context of interstellar chemistry discussed in the next chapter.

2.2.2 Reactions on the Surfaces of Interstellar Dust Grains

The difficulties of making interstellar H_2 with adequate efficiency were already apparent several decades ago, and—more or less by default—astronomers began to consider the possibility that hydrogen atoms could recombine efficiently to form hydrogen molecules on the surfaces of interstellar dust grains. The idea was a simple one: a H-atom arrives at the surface, sticks to the surface in a weak binding site (so-called physisorption) and shares the surface temperature, diffuses over the surface until it locates another weakly-bound H-atom and forms a bond with it; the newly formed H_2 molecule may desorb immediately into the gas phase or in some later event. Chemists call such a sequence of events a Langmuir–Hinshelwood reaction. The mobility depends on the surface temperature and at the lowest temperatures H-atoms may be mobile because of quantum mechanical tunnelling through the energy barriers between binding sites. When the two atoms meet at the same site they interact to form H_2, releasing the binding energy of the H_2 molecule, 4.5 eV. Some of this

Interstellar Molecular Hydrogen

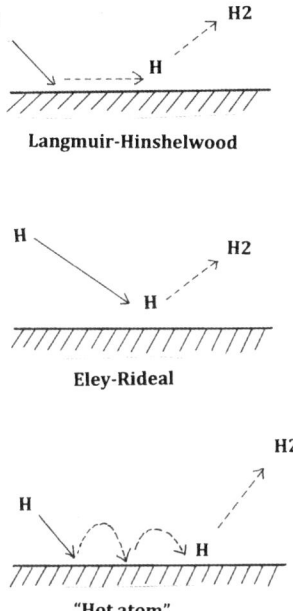

Figure 2.4 Schematic diagrams illustrating the Langmuir–Hinshelwood, Eley–Rideal, and 'hot-atom' mechanisms for forming molecular hydrogen on a surface.

energy is shared with the surface while the rest is retained as kinetic, vibrational and rotational energies in the ejected molecule. A schematic diagram illustrating this Langmuir–Hinshelwood process is shown in Figure **2.4**.

Results from simple models of this process were compared to the H_2 formation rate inferred from the observed H_2 abundance and an estimated loss rate for the photodissociation of H_2 by the local radiation field (see Section **2.3.1**). While there were many uncertainties in the model and in the inferred interstellar loss rate, nevertheless the picture was encouraging. It required that H-atoms should stick efficiently on dust grain surfaces, and that the grain temperatures should be low. If the grain temperature was too high, then the first H-atom would have evaporated before it could locate the second. If the grain temperature was too low, then the H-atom might have too low a mobility over the surface to locate another H-atom before it desorbs. But within these constraints, then the picture suggested that almost every H-atom arriving at the surface of a dust grain should leave as part of an

H_2 molecule. The newly formed molecule would be hotter (*i.e.* moving faster) than the local H_2 molecules, and would be vibrationally and rotationally excited.

In reality, the surface is likely to be far from the perfect lattice envisaged in the simple model. The surface must include many types of dislocations and defects, and perfect structures may be rather rare and not extensive. So an incident H-atom might experience a range of binding energies. The largest binding energies are associated with strong chemical bonds, *i.e.* chemisorption, rather than physisorption. When this was included in the models, the range of dust grain temperatures over which H_2 molecules could form was increased, making the formation mechanism more plausible for interstellar medium conditions.

There the matter largely rested until the 1990s, when techniques in the laboratory improved so much that fairly realistic experiments could be performed representing interstellar surface chemistry. At the same time, computing power also improved dramatically, so that detailed and reliable quantum mechanical calculations of H_2 formation could be performed. In general, these computations have confirmed that the sticking of H-atoms to surfaces should be efficient, and have determined the nature of the atom−surface bond. These theoretical results have helped to interpret the data from the experiments. Important laboratory studies of H_2 formation have been made in the USA, in the UK, in Denmark, and in France. We shall discuss some of this work in more detail, in Chapter **9**.

It now seems clear from such experiments that different mechanisms may operate, depending on the experimental conditions and the nature of the surface. When the surface temperature is very low and the atom fluxes not too large, then H_2 formation occurs by the Langmuir−Hinshelwood mechanism. But this mechanism does give a rather narrow range for efficient H_2 formation. At higher surface temperatures, another mechanism begins to operate, extending the temperature range for efficient H_2 formation. This second mechanism involves a chemisorbed H-atom interacting directly with a gaseous H-atom to form a molecule; chemists call this mechanism an Eley−Rideal reaction (see Figure **2.4**). In fact, the process may be rather more complicated.

It is also possible that the gaseous H-atom interacts with the surface and undergoes a number of collisions with it, gradually losing energy until it sticks to the surface and shares its temperature. During these interactions, the atom is moving faster than would be expected for the temperature of the surface. If it interacts with a chemisorbed H-atom it may form H_2 quite efficiently. This is called the 'hot atom' mechanism (see Figure **2.4**). It is more likely to occur when the surface coverage by adsorbed H-atoms is high.

The energy state of the newly formed molecule is important. The kinetic energy can be shared with other gas phase atoms and molecules, and represents a heat source for the interstellar gas. Detection of emission from vibrationally and rotationally excited states of newly formed H_2 could give a direct measure of the H_2 formation process in the interstellar medium. Recent experiments suggest that perhaps 40% of the binding energy of a newly-formed H_2 molecule (*i.e.* nearly 2 eV) may be injected into the dust surface, a considerable heat source for a dust grain.

Of course, to assess whether this formation process is significant in the interstellar medium, one needs to make some assumptions about the size distribution and total number of dust grains, and their nature. But with conventional ideas about grains, supported by observations and theory, it seems safe to conclude from work over the last two decades that the formation of H_2 on a variety of astrophysically plausible surfaces, as measured in the laboratory, can be an efficient process over a reasonably wide range of surface temperatures. Depending on the physical conditions in the experiment (or in the interstellar medium) the process may involve Langmuir–Hinshelwood, Eley–Rideal, and 'hot atom' reactions. Adopting current ideas concerning the size distribution and composition of interstellar dust grains, then the implication is that nearly every H-atom arriving at a grain must stick to the surface until it reacts and leaves as part of an H_2 molecule.

2.2.3 Wilder Speculations

A variety of laboratory experiments show the rather surprising result that solids such as amorphous carbon at low temperature can retain a population of radicals. Raising the temperature of such solids by a small amount, or increasing the population of radicals, causes the radicals to recombine explosively in a runaway event, with associated emission of visible and near infrared radiation. Evidently, the solid in these experiments is raised to a high temperature (typically on the order of a thousand K) by the sudden release of chemical energy during the recombination of the radicals.

It is well known that amorphous carbon is capable of trapping hydrogen atoms at interstitial and internal sites with binding energies that are large enough restrict the mobility. Therefore, it seems possible that in the hydrogen-rich interstellar medium cold interstellar grains of amorphous carbon may accumulate significant populations of atomic hydrogen. The laboratory experiments suggest that a minor temperature fluctuation of the grain (or an increase in the radical population to a critical level) may be enough to trigger a runaway recombination of the atomic hydrogen to form H_2 molecules, with an almost instantaneous and significant temperature rise in the interstellar grains and in the ejected hydrogen molecules.

On this picture, molecular hydrogen formation does not depend on the delicate balance of binding energy and temperature, giving a rather narrow temperature range under which H_2 formation occurs. Instead, the runaway recombination of the H-atoms triggers a very significant temperature rise in the dust and the molecules. There are several important consequences; firstly, the grain temperature is raised to such a level that it can radiate in the infrared; secondly, the H_2 molecules formed in this process arise intermittently, rather than continuously, and finally, the newly-formed H_2 molecules—and any other H_2 molecules residing on the grain—are abruptly desorbed and share a common high temperature on the order of a thousand K.

It remains to be seen whether this process, quite different from the Langmuir–Hinshelwood, Eley–Rideal, or 'hot-atom' mechanisms, will be confirmed in further work and whether it will contribute to the formation of molecular hydrogen in the interstellar medium. But this mechanism would have the advantage that H_2 formation could occur not only at the relatively low temperatures of interstellar clouds, 10–100 K, but also in much warmer regions close to stars. It appears from observations that H_2 is required to be formed in these warmer near-stellar regions.

2.3 HOW IS INTERSTELLAR MOLECULAR HYDROGEN DESTROYED?

Interstellar molecular hydrogen can be destroyed by ultraviolet radiation, by cosmic rays, or in reactions with other species. The dominant destruction mechanism depends on the environment.

2.3.1 Photodestruction in the Interstellar Radiation Field

The radiation field impinging on interstellar clouds originates in massive stars with surface temperatures of tens of thousands of K, compared to the more modest surface temperature of the Sun; about 6000 K. Sunlight has maximum intensity at a colour that is yellowish white. The much hotter stars emit radiation that peaks at wavelengths that are much shorter, typically well into the ultraviolet, near 100 nanometres. However, not all of this radiation impinges on interstellar clouds.

Each massive star is surrounded by hydrogen, which absorbs ultraviolet light and becomes ionised. The protons and electrons produced by this ionisation recombine to make neutral atoms which—again—are ionised by the radiation field. In effect, the star is surrounded by a fairly sharp-edged sphere of ionised hydrogen that absorbs all the ultraviolet radiation—from short wavelengths up to the wavelength capable of ionising hydrogen, which is in fact 91.2 nanometres (nm). Wavelengths longer than this escape from the ionised sphere of gas, and ultimately impinge on interstellar clouds. Figure **2.5** illustrates this schematically and Figure **2.6** shows a galaxy with many HII regions, some roughly spherical and with fairly sharp boundaries. At a typical point in a galaxy, the intensity of starlight will have contributions from a number of hot stars.

So the radiation field in an interstellar cloud has the wavelength dependence of radiation from a hot star, extending from 91.2 nm in the ultraviolet through the visible and into the infrared, but with zero intensity for wavelengths in the ultraviolet shorter than 91.2 nm. What effect can this radiation field have on H_2?

Radiation can directly dissociate H_2 molecules:

$$H_2 + radiation \rightarrow H + H$$

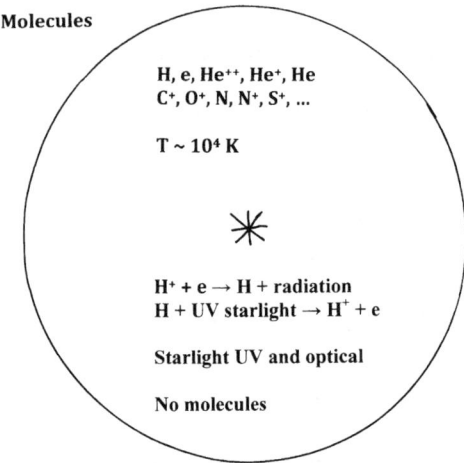

Figure 2.5 A schematic diagram illustrating the region of ionised hydrogen around a hot star.

(where one of the H-atoms is electronically excited) but wavelengths shorter than about 85 nm are required to make this 'photodissociation' occur, and we have seen that radiation in this wavelength range is not available in interstellar clouds. Photodissociation into two H-atoms both of which are in their ground electronic states only requires radiation shorter than about 280 nm and this radiation is available; however, the transition is very strongly forbidden by quantum mechanics and does not occur.

Another possibility is that radiation may ionise H_2. This 'photoionization' represents a loss of H_2, since the H_2^+ ion may go on to react with other species, for example

$$H_2 + \text{radiation} \rightarrow H_2^+ + e$$

Figure 2.6 The galaxy NGC 2403 has many well-defined ionised hydrogen regions around hot stars (Image credit: Subaru Telescope (NAOJ), Hubble Legacy Archive, processing Robert Gendler).

$$H_2^+ + e \rightarrow H + H$$

the latter reaction being an example of 'dissociative recombination', since the electron recombines with the ion but the resulting neutral species falls apart, *i.e.* dissociates. However, the wavelength of the radiation required to ionise H_2 must be shorter than about 80 nm, and is not available in interstellar clouds.

So neither of the obvious processes, direct photodissociation or photoionisation, can destroy H_2 in interstellar clouds. Does this mean that H_2, once formed in an interstellar cloud, exists forever?

Well, no! There is a more subtle process that destroys H_2, using radiation that *is* available in interstellar clouds. This mechanism depends on the fact that H_2 is a molecule with vibrational modes. Imagine that we start with the H_2 molecule in its ground electronic and lowest vibrational states. Radiation with a wavelength, λ, of about 110 nm excites the molecule into an upper electronic state, and into several vibrational levels of that upper state. (In fact, these are the transitions by which interstellar H_2 was first observed in absorption; see Section **2.1**, above.) The excited molecule quickly relaxes back to the ground electronic state, but it ends up in a distribution of vibrational levels, v'', some in the lowest level, $v'' = 0$, some in $v'' = 1$, $v'' = 2$, *etc.*, and some in the highest possible bound vibrational level $v'' = 14$. If the molecule is given any more vibrational energy than represented by this state, it will fall apart. That is exactly what happens in this relaxation to the ground state. In most cases, the molecule remains bound, but in a significant number of excitations, around 20%, the molecule falls apart:

$$H_2 + \text{radiation } (\lambda \sim 110 \text{ nm}) \rightarrow H + H$$

The process is illustrated schematically in Figure **2.7**.

Note that the initial excitation that leads to dissociation is from one particular electronic + vibrational + rotational state to another. This means that a very specific wavelength of radiation is required. So as the radiation penetrates an interstellar cloud, the radiation of the specific wavelength is used up, so the chance of

Interstellar Molecular Hydrogen

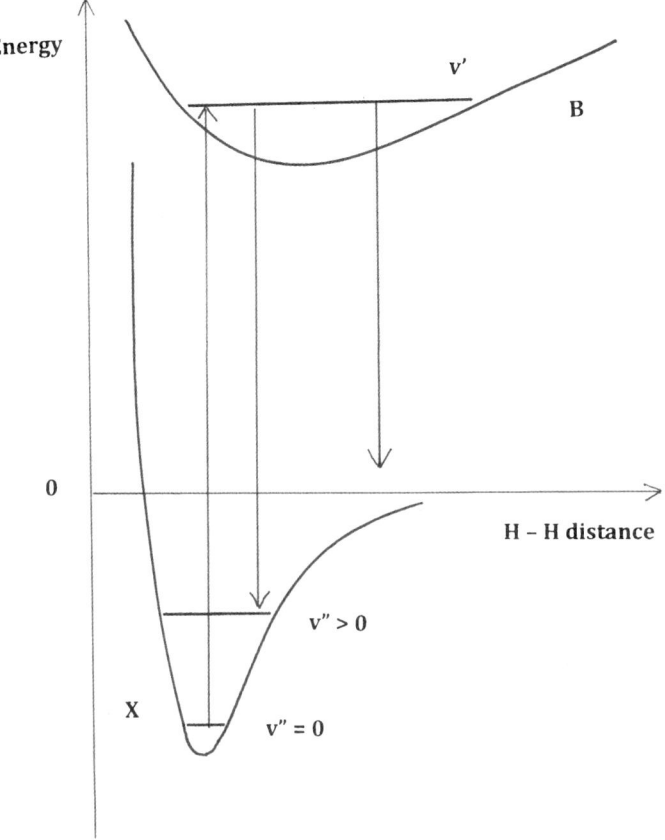

Figure 2.7 A schematic diagram illustrating the photodissociation of interstellar molecular hydrogen.

destroying H_2 inside a cloud is much less than at the boundary. At the boundary, an H_2 molecule may survive for (only) a thousand years or so, while in the interior of a diffuse cloud a molecule may survive for a period approaching a million years.

2.3.2 Ionisation and Dissociation of H_2 by Cosmic Rays

Cosmic rays are fast particles, mostly protons, helium nuclei and electrons. Their effects on interstellar gas are to ionise and heat it, and the most effective cosmic rays are those with energies of a few MeV. These are such high energies that the loss of a few tens of electron volts in ionising an atom is negligible, so we do not have to account for the decay of the energy of cosmic rays, except in massively dense objects.

Cosmic rays (cr) destroy H_2 by dissociating and ionising the molecule. There are various channels:

$$H_2 + cr \rightarrow H_2^+ + e + cr$$

$$H_2 + cr \rightarrow H + H + cr$$

$$H_2 + cr \rightarrow H^+ + H + e + cr$$

$$H_2 + cr \rightarrow H^+ + H^- + cr$$

The first, 'collisional ionisation', has the most rapid rate. If the flux of cosmic rays has the value believed to be the case for the Milky Way, then a hydrogen molecule may on average be ionised once every few billion years. As a destruction mechanism for H_2, this is obviously very much less effective than photodestruction (see Section **2.3.1**, above).

The second channel, 'collisional dissociation' has a rate that is roughly a factor of ten lower than the first channel, while the third channel, 'collisional dissociative ionisation', is about another factor of ten smaller. The last channel, another form of dissociative ionisation, has a rate that is a factor of about a hundred smaller than the third channel.

The main loss route for H_2 is clearly the first one, forming H_2^+. But what happens to the newly formed H_2^+? Could it be returned to H_2? If so, the reaction wouldn't be a loss mechanism. In fact, the importance of cosmic ray ionisation of H_2 is greatest in denser, darker regions of the interstellar medium, from which starlight is largely excluded. Because that is the case, most of the hydrogen is already in molecules so the most likely collision that can occur with H_2+ is with another H_2:

Interstellar Molecular Hydrogen

$$H_2^+ + H_2 \rightarrow H_3^+ + H$$

where H_3^+ is a new molecular ion, protonated molecular hydrogen, a stable triangle of protons with two attendant electrons. It's a detected interstellar molecule and a very important species, as we will see in Chapter 3, because it can donate a proton very efficiently to other species, *e.g.*

$$H_3^+ + CO \rightarrow H_2 + HCO^+$$

This example of a 'proton donation' converts CO into the formyl radical ion (HCO^+), another observed species and important tracer of dense gas. Proton donation is at the heart of gas phase astrochemistry, as we'll discuss in the next chapter. In this reaction, we recover one hydrogen molecule, but the original hydrogen molecule ionised by the cosmic ray has indeed been truly lost, one H-atom ending in the gas phase, and the other (in this example) ending as part of the molecular ion HCO^+.

If H_3^+ does not react by donating a proton, it may meet an electron

$$H_3^+ + e \rightarrow H_2 + H$$

and react in a 'dissociative recombination'; such a recombination is generally very fast, occurring on almost every collision. The outcome of a cosmic ray ionisation of H_2 is just a question of whether an H_3^+ molecule meets a CO (or other) molecule or an electron. It is clear, however, that each cosmic ray ionisation of H_2 to form an H_2^+ ion and subsequent dissociative recombination does effectively destroy one H_2. Subsequent ion-molecule reactions may destroy even more hydrogen molecules (see Section **2.5**). The other channels also destroy one H_2 per collision, but less effectively.

2.3.3 Reactions of H_2 with He^+

Helium is the only element that has an abundance close to that of hydrogen (see Chapter **1**, Table **1.1**). Of course, neither H nor H_2

reacts directly with the noble element, He. However, helium is subject to ionization by cosmic rays, and reactions with He^+ can be very destructive. The ionisation potential of helium is 24.59 eV, so when the helium ion meets another atom or molecule with a lower ionisation potential it grabs an electron at every opportunity. The rates of the following reactions:

$$H_2 + He^+ \rightarrow H_2^+ + He$$

$$H_2 + He^+ \rightarrow H^+ + H + He$$

have been measured in the laboratory, and the reactions are relatively slow. The first reaction (charge transfer) occurs in about one collision in 10^5 while the second (dissociative charge transfer) is a few times faster. These low rates are, in fact, to be expected for very fast atomic collisions.

In most situations in the interstellar medium the abundance of He^+ is low, and so the rate of loss of H_2 by reactions with He^+ is usually much slower than the direct cosmic ray ionisation of H_2.

2.4 THE H/H_2 BALANCE IN INTERSTELLAR CLOUDS OF THE MILKY WAY GALAXY

In the interior of dark interstellar clouds where the total hydrogen density is more than one thousand H-atoms per cm^3 (see Chapter 1, Table **1.2**), the main formation mechanism for H_2 is in reactions of atomic hydrogen on dust grains, and the destruction is by cosmic rays. The actual number density of free H-atoms in dark clouds can be computed for Milky Way galaxy conditions to be very roughly about one H-atom per cm^3, if the chemistry has reached steady state. So in these clouds the hydrogen is almost entirely molecular.

The situation is different in diffuse clouds where the destruction of H_2 molecules occurs by absorption of starlight. As pointed out in Section **2.3.1**, this process depends on absorption of radiation in a narrow line, and can run out of radiation for greater depths within the cloud. In other words, the outer layers of a cloud absorb

Interstellar Molecular Hydrogen

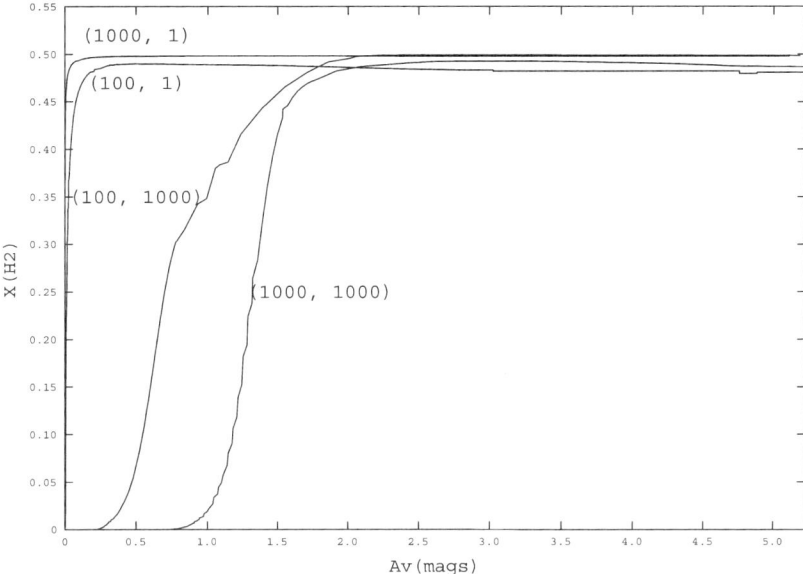

Figure 2.8 The conversion of atomic hydrogen to hydrogen molecules, computed as a function of depth into an interstellar cloud (Courtesy: S. Viti) X is the fraction of all the hydrogen atoms converted to H_2. Each curve is labelled with two parameters: the first is the number density of all hydrogen atoms (in both H and H_2) per cm^3, and the second is the intensity of UV radiation assumed to impinge on the cloud, in units of the mean interstellar radiation field of the Milky Way galaxy. The curves are shown as a function of visual extinction caused by interstellar dust, A_V, measured from the cloud edge. Very roughly, one magnitude of visual extinction is equal to one optical depth in the visual.

the radiation that would otherwise destroy the hydrogen molecules in the interior. This is called 'self-shielding'. The balance of formation and destruction gives little H_2 near the edge, but a lot in the interior. It turns out that the transition between mostly atomic hydrogen and mostly molecular hydrogen is fairly sharp, and generally occurs close to the edge of even a low-density cloud (see Figure **2.8**).

Therefore, in our considerations of interstellar chemistry, where H_2 plays a very important role, we do not need to worry too much about the narrow boundary region of interstellar clouds where the hydrogen is mainly atomic.

2.5 WHY IS MOLECULAR HYDROGEN IMPORTANT IN INTERSTELLAR CHEMISTRY?

Molecular hydrogen is much more abundant in the interstellar medium than any other atom or molecule with which it can react. Of course, it does not react with helium atoms, which are nearly comparable in abundance, but it may under appropriate conditions react with atoms or ions of oxygen, carbon, nitrogen, sulfur, and all the other elements, or with molecules or molecular ions containing them. This preponderance of hydrogen means that any process that can occur with H_2 or its derivatives is almost certain to be important for astrochemistry. Generally, we can assume that if a species can react with H_2, then this happens much more rapidly than any other reaction. It is only when a species can react no further, that other routes become important. For example, the oxygen atomic ion, O^+, can react with H_2 in an ion-molecule exchange:

$$O^+ + H_2 \rightarrow OH^+ + H$$

on almost every collision. The product, OH^+, also reacts very easily with H_2:

$$OH^+ + H_2 \rightarrow H_2O^+ + H$$

to form a new product, H_2O^+, which can undergo one further reaction with H_2, and this occurs on almost every collision:

$$H_2O^+ + H_2 \rightarrow H_3O^+ + H$$

(Note that this sequence of reactions successively destroys H_2 molecules). However, this hydrogen addition process stops with H_3O^+, as the oxygen ion now shares the additional three electrons necessary to complete its outer electronic shell. The next stage, therefore, involves some other less frequent collisions; for example, the hydronium ion (H_3O^+) may dissociatively recombine with an electron to give a variety of products:

$$H_3O^+ + e \to H_2O + H$$

$$\text{or } OH + H + H$$

$$\text{or } O + H_2 + H$$

$$\text{or } OH + H_2$$

or may react with neutral atoms, such as carbon

$$H_3O^+ + C \to HCO^+ + H_2$$

or interact with a low ionisation potential species, such as sodium, in a dissociative charge exchange

$$H_3O^+ + Na \to H_2O + H + Na^+$$

All these reactions give product molecules that have been detected in the interstellar medium. Note that these routes depend entirely on the presence of H_2; without H_2 these schemes cannot operate. We shall discuss all these types of process in Chapter **3**.

The variety of processes in which molecular hydrogen can take part is immense; those that we have discussed here are listed in Table **2.1**.

These reactions include almost all types of process commonly assumed to occur in the interstellar medium. Our discussion of interstellar molecular hydrogen has therefore given a good introduction to the contents of Chapter **3**. The only significant reaction type that is missing from this list in Table **2.1** is solid-state

Table 2.1 Lists all the types of reaction that we have mentioned in this chapter, with an example (cr represents cosmic ray, rad represents radiation, and g represents a dust grain).

Type of reaction	Example involving hydrogen
Associative detachment	$H^- + H \rightarrow H_2 + e$
Chemical exchange	$OH + H \rightarrow H_2 + O$
Collisional dissociation	$H_2 + cr \rightarrow H + H + cr$
Collisional dissociative ionisation	$H_2 + cr \rightarrow H + H^+ + e + cr$
Collisional ionisation	$H_2 + cr \rightarrow H_2^+ + e + cr$
Dissociative recombination	$H_2^+ + e \rightarrow H + H$
Ion–molecule reaction	$H_2^+ + H \rightarrow H_2 + H^+$
Photodetachment reaction	$H^- + rad \rightarrow H + e$
Photodissociation reaction	$H_2^+ + rad \rightarrow H + H^+$
Photoionisation reaction	$H_2 + rad \rightarrow H_2^+ + e$
Proton donation reaction	$H_3^+ + CO \rightarrow H_2 + HCO^+$
Radiative association	$H + O \rightarrow OH + rad$
Surface reaction	$H + H + g \rightarrow H_2 + g$
Three-body reaction	$H + H + H \rightarrow H_2 + H$

chemistry that may occur in ices that accumulate on interstellar dust grains inside dark clouds. We shall discuss that topic in Chapter **3**.

FURTHER READING

T. W. Hartquist and D. A. Williams, *The Chemically Controlled Cosmos*, Cambridge University Press, UK, 1995.

F. Combes and G. Pineau des Forêts, eds, *Molecular Hydrogen in Space*, Cambridge Contemporary Astrophysics, Cambridge University Press, UK, 2000 (conference proceedings).

CHAPTER 3
Chemical Routes to Interstellar Molecules

3.1 MAKING INTERSTELLAR MOLECULES

In this section, we will look briefly at various routes by which interstellar molecules might be formed. In subsequent sections, we shall examine each of these routes in more detail.

3.1.1 Radiative Associations of Atoms

Diffuse interstellar clouds are relatively simple, and can provide useful tests of interstellar chemistry. The abundances of atomic and molecular hydrogen have been measured in many diffuse clouds, and the relative abundances of the chemically important elements oxygen, carbon and nitrogen may be assumed to be solar. So the potential ingredients for diffuse cloud chemistry are understood. To begin with, therefore, we will discuss, in this subsection, some problems of making molecules in diffuse clouds.

It is easy to show that the simple collision of, say, an oxygen atom with a hydrogen atom to make hydroxyl (OH), or of a carbon ion with a hydrogen atom to make the methylidyne ion (CH^+), cannot account even for the rather small abundances of these molecular species observed in diffuse clouds. Typically, the abundances are around 10^{-8} by number, relative to H-atoms. If the addition of one atom to another is to make a stable molecule, the colliding pair must lose some energy. In diffuse clouds, the number density in the gas is far too low for a third atom to have a good chance of impacting on the colliding pair and removing some energy. So the only possibility of forming a molecule is that the colliding pair emits some radiation, stabilising the pair, as implied in these reactions:

$$O + H \rightarrow OH^* \rightarrow OH + \text{radiation}$$

$$C^+ + H \rightarrow CH^{+*} \rightarrow CH^+ + \text{radiation}$$

where the asterisk (*) indicates that the colliding pair have sufficient energy to dissociate, and the emitted radiation carries away some energy from the colliding pair. Once that radiation has been emitted, the pair no longer has enough energy to separate. The rate coefficients for each of these radiative association reactions can be computed fairly accurately using quantum mechanics. Their values are small, implying that in a gas at a temperature less than a few hundred K (appropriate for diffuse clouds) very roughly only one in about ten million collisions lead to some radiation being emitted. Therefore, only a tiny fraction of all such collisions can form a molecule. When one takes into account the loss of molecules such as OH and CH^+ in photodissociation by ambient starlight, or by other reactions such as dissociative recombination of CH^+ with electrons:

$$OH + \text{starlight} \rightarrow O + H$$

$$CH^+ + e \rightarrow C + H$$

then one finds that these routes fail to make OH and CH^+ in the observed amounts by factors of a hundred or more. Now, astrochemistry can be a fairly crude business but factors of a hundred are definitely too large to ignore! We need to determine whether much faster gas phase chemical routes exist for the formation of these and other species.

3.1.2 Exchange reactions

If direct associations of atoms or atomic ions with atomic hydrogen cannot do the job, then exchange reactions with molecular hydrogen are the next most obvious candidate to consider. As we have seen in Chapter **2**, molecular hydrogen exists in abundance in many diffuse clouds. Is the reaction:

$$O + H_2 \rightarrow OH + H$$

a suitable route to forming OH in diffuse clouds? For this reaction to proceed, then—at a minimum—the bond between the H-atoms (4.5 eV) in H_2 must be broken while at least some of the energy in forming the OH bond (4.4 eV) must be released. Evidently, the energy released is less than the energy required, so the reaction will not proceed unless assisted in some way. Clearly, there is a significant barrier to the reaction occurring, and experiments show that temperatures around 1000 K or more are required to make the efficiency reasonably high.

There is a similar story for the reaction C^+ ions with H_2 to form CH^+; molecular hydrogen is more strongly bound than the methylidyne ion. Energy in the form of heat must be inserted to make this reaction proceed, and the simplest way to do this is to raise the temperature in the gas of H_2 and C^+ to more than 1000 K. The reactions of nitrogen and sulfur atoms with molecular hydrogen are similarly inhibited at low temperatures.

Evidently, rather high temperatures are required to drive reactions such as those of O-atoms and C^+-ions with H_2 molecules. Temperatures of thousands of K are not expected in diffuse clouds, where temperatures around 100 K are normally inferred. Evidently, heat can drive some reactions that would not otherwise occur. We will discuss this process in more detail in Section 3.5.

3.1.3 Radiative Associations with Molecules

However, all is not lost for radiative association to be a route to interstellar chemistry. Consider binary radiative associations involving molecules. It is easy to see that for colliding systems more complicated than two atoms (as discussed for OH and CH^+ in Section **3.1.1**), the complex formed in the collision should last longer and therefore give a greater chance that radiation can be emitted. If a complex has several atoms, then there are many possible modes of vibration, and the excess energy that can—potentially—break a bond in the complex is generally shared among all these modes and only rarely appears in one single mode where it causes dissociation. A complex larger than diatomic can, therefore, have a long lifetime. For example, the radiative association of CH_3^+ and H_2O:

$$CH_3^+ + H_2O \rightarrow CH_3OH_2^+ + \text{radiation}$$

to form protonated methanol occurs in about one in a hundred collisions. This is a much more favourable situation than the very rare chance of about one in ten million for stabilising a diatomic collision complex like OH*. Still larger collision complexes may have lifetimes so long that almost every one of them can be radiatively stabilised. In such a case, radiative association occurs on almost every collision.

Even changing from a diatomic collision complex to a triatomic collision complex makes a big difference to the radiative association rate. For example, the radiative association:

$$C^+ + H_2 \rightarrow CH_2^+ + \text{radiation}$$

has a rate coefficient that is about a hundred times larger than that of the radiative association between C^+ and H. However, it is still a slow reaction, occurring only once in about a hundred thousand collisions. After formation, CH_2^+ reacts very rapidly with electrons:

$$CH_2^+ + e \rightarrow CH + H$$

forming the methylidyne molecule. This radiative association of C^+ with H_2 is still inefficient; nevertheless, it is sufficiently fast that it is thought to be responsible for the formation of CH in diffuse clouds. The association is faster than similar associations between neutral species, because one of the partners is the ion, C^+, which induces a long-range interaction in the other, H_2, distorting the charge distribution in the molecule and creating a dipole; this ion–dipole interaction helps to draw the reactants together. So ions can help interstellar chemistry to proceed.

3.1.4 Ionisation by Starlight

We have seen from this discussion of CH formation that it is beneficial to have ions, because of the long-range interactions that they induce in their collisions with other species. But why is carbon ionised in diffuse clouds? After all, the neutral carbon atom is in a lower energy state than the ion by energy equal to the ionisation potential of carbon, 11.2 eV. Which other species are present as atomic ions in diffuse clouds?

Carbon atoms are photoionised by the starlight that penetrates diffuse clouds. These clouds have only a modest amount of dust in them, so the radiation of the average interstellar radiation field (dominated by emission from nearby bright stars) suffers only a slight reduction in intensity. Thus, starlight puts energy into the gas by ionising carbon (and some other elements) and enabling the radiative association of C^+ with H_2 to occur sufficiently rapidly to make a significant contribution to the chemistry. Without starlight (or other sources of ionisation), there would be no carbon ions and the association reaction of those ions with molecular hydrogen would not occur. Thus, starlight helps to promote interstellar chemistry. We shall discuss the importance of starlight and other forms of electromagnetic radiation in Section **3.2**. The energy input from starlight in diffuse clouds liberates electrons with excess energy that is shared with the gas in collisions; this is the main reason why diffuse clouds have temperatures around 100 K, rather warm compared to darker clouds, where the temperature is typically around 10 K.

3.1.5 Ionisation by Cosmic Rays

Cosmic rays pervade the entire volume of interstellar space. As we saw in Chapter **2**, cosmic rays are fast particles, mostly protons and electrons, and helium ions. They range widely in energy, but those that interact most effectively with the gas have energies of a few MeV up to a few hundred MeV. Cosmic rays (c.r.) of such high energies are obviously capable of ionising all interstellar species and, as discussed in Section **2.3.2**, will ionise molecular hydrogen, for which the most likely channel leads to the hydrogen molecular ion (H_2^+):

$$H_2 + c.r. \rightarrow H_2^+ + e + c.r.$$

In gases where much of the hydrogen is molecular, then the most likely reaction for the hydrogen molecular ion is with another hydrogen molecule to form protonated molecular hydrogen (H_3^+), a stable molecule of triangular form. H_2 has a rather low proton affinity (*i.e.* a low binding energy between H_2 and H^+). See Table **3.1** for proton affinities of some relevant species. H_3^+ readily donates a proton to almost all other interstellar species.

In Section **2.3.2**, we saw that H_3^+ can convert carbon monoxide (CO) to the formyl radical ion (HCO^+), an important interstellar species. In another example, H_3^+ can donate a proton to an oxygen atom (on almost every collision):

$$H_3^+ + O \rightarrow H_2 + OH^+$$

Table 3.1 Proton affinities of some molecules of astronomical interest. Note that molecular hydrogen has a lower proton affinity than any other molecule listed, apart from molecular oxygen. This implies that protonated molecular hydrogen, H_3^+, is able to transfer a proton to any of the molecules listed here, other than O_2.

Molecule	Proton affinity (eV)	Molecule	Proton affinity (eV)
H_2	4.42	CH_4	5.57
CO	6.19	H_2O	7.20
N_2	5.15	NH_3	8.90
NO	5.53	H_2CO	7.43
O_2	4.38	HCN	7.43
SH	7.19	CH_3OH	7.85
CS	7.63	CO_2	5.64
HCl	5.88	OCS	6.58

and the OH$^+$ ion provides an entry to subsequent chemistry. For example, it can rapidly undergo successive reactions with hydrogen molecules (the most likely interaction, given the dominance of hydrogen in the interstellar medium), abstracting a hydrogen atom in each reaction:

$$OH^+ \rightarrow H_2O^+ \rightarrow H_3O^+$$

until the triple valency of O$^+$ is satisfied at the stage of H$_3$O$^+$ (note that this is a detected interstellar species). No further additions of H are possible after that stage, and the most likely next step is that the protonated water (H$_3$O$^+$) dissociatively recombines with an electron to form water:

$$H_3O^+ + e \rightarrow H_2O + H$$

Many such schemes are initiated by reaction with H$_3{}^+$ and subsequent ion−molecule reactions. Thus, cosmic ray ionisation of molecular hydrogen is an important driver of the chemistry—particularly in dark clouds where starlight is excluded by the extinction due to interstellar dust. We will discuss this important driver of interstellar chemistry in more detail below, in Section **3.3**.

3.1.6 Dust Grains

The most effective way of forming molecular hydrogen in interstellar clouds is—as discussed in Chapter **2**—in surface reactions on interstellar dust grains (if they are present, mixed with the gas, as they usually are). There are strong indications that similar hydrogenation reactions may occur for oxygen and nitrogen atoms, and possibly also for carbon atoms. But, regardless of hydrogenation reactions, insofar as almost all interstellar chemistry depends on the presence of molecular hydrogen, then very nearly all interstellar chemistry may be regarded as being driven by surface reactions. Clearly, the presence of dust grains opens up important routes to interstellar chemistry.

There are two different ways in which dust grains contribute actively to interstellar chemistry—apart from the passive shielding of molecules from the destructive effects of the interstellar radiation field, helping them to survive. The first is important in interstellar regions rich in atomic hydrogen and involves surface reactions forming molecular hydrogen and possibly other simple species such as water (H_2O), ammonia (NH_3), methane (CH_4), and hydrogen sulfide (H_2S). The second active contribution is in the formation and processing of ices that accumulate on the surface of dust grains in cold and almost entirely molecular dark clouds. We shall see that while ices as originally deposited on grains are chemically rather simple, the chemical processing that they undergo leads to some of the greatest chemical complexity that is found in interstellar clouds and especially in star-forming regions. We will discuss the roles of dust grains as promoters of interstellar chemistry in more detail below, in Section **3.4**.

3.2 ELECTROMAGNETIC RADIATION AND INTERSTELLAR CHEMISTRY

Though photons remove some species, they also drive the chemistry in many astronomical environments by creating ions that react with molecules.

3.2.1 Starlight

What is the chemistry promoted by starlight in interstellar clouds? We will assume that the clouds are cold, that hydrogen is present in both atomic and molecular forms, and that starlight is the only factor promoting chemistry.

What can the starlight do? That depends on the range of wavelengths in the spectrum of the starlight. The radiation from massive stars extends from the infrared into the far ultraviolet. However, given the predominance of hydrogen in the interstellar medium, all radiation capable of ionising atomic hydrogen is confined to sharp-edged ionised regions surrounding stars (see Figure **2.9**), and so stellar radiation penetrating clouds beyond these ionised regions contains only those wavelengths longer than the limit for ionising hydrogen, *i.e.* 91.2 nanometres in the far ultraviolet, and has zero intensity at shorter wavelengths. Even with this cut-off, the radiation field with wavelengths longer than 91 nm is still capable of ionising the fairly abundant elements of carbon, sulfur, chlorine, and many metals such as silicon, iron, magnesium, sodium and potassium (although most of these metal atoms are locked up in dust grains and do not play much of a role in gas phase chemistry). See Table **3.2** for ionisation potentials of some relevant atoms.

So in diffuse clouds where dust grains do not absorb all the starlight we expect the elements carbon, sulfur and chlorine to be ionised. But the abundant elements oxygen and nitrogen are present as neutral atoms because their ionisation potentials are equal to or larger than that of hydrogen. Therefore, starlight pervading interstellar clouds cannot ionise oxygen and nitrogen atoms, and does not drive a direct chemistry of oxygen, nitrogen, or sulfur since exchange reactions of O, N, and S atoms with hydrogen molecules are suppressed at low temperatures (*cf.* Section **3.1.1**).

However, as we have seen in Section **3.1.3**, the radiative association of carbon ions—created by starlight—with molecular hydrogen leads to the formation of the methylidyne radical (CH). This radical can then take part in barrier-free neutral exchange reactions to form new species, for example, in reaction with O-atoms to form carbon monoxide:

$$CH + O \rightarrow CO + H$$

with N-atoms to form the cyanide radical:

$$CH + N \rightarrow CN + H$$

Table 3.2 Ionisation potentials (I.P. in eV) of some species of astrochemical interest.

Atom	I.P.	Atom	I.P.	Atom	I.P.	Atom	I.P.	Molecule	I.P.
H	13.60	O	13.62	Si	8.15	K^+	31.63	H_2	15.43
He	24.59	Na	5.14	Si^+	16.35	Ca	6.11	C_2	12.00
C	11.26	Na^+	47.29	S	10.36	Ca^+	11.87	CN	13.80
C^+	24.38	Mg	7.65	S^+	23.34	Fe	7.90	CO	14.01
N	14.53	Mg^+	15.04	K	4.34	Fe^+	16.19	CH_4	12.60

or with S to form CS. These new species are detected in interstellar diffuse clouds, and elsewhere in the interstellar medium.

Sulfur is present in diffuse clouds as S^+, but—like C^+ ions—S^+ ions do not react directly with hydrogen molecules, although—like C^+ ions—they can radiatively associate (rather slowly) with H_2 to form H_2S^+:

$$S^+ + H_2 \rightarrow H_2S^+ + \text{radiation}$$

Dissociative recombination then forms sulfur monohydride, SH, a detected interstellar species:

$$H_2S^+ + e \rightarrow SH + H$$

Another route to sulfur-bearing species occurs when sulfur ions and atoms react in exchange reactions with CH to form CS, an important detected interstellar species.

However, chlorine ions can react directly and efficiently with H_2:

$$Cl^+ + H_2 \rightarrow HCl^+ + H$$

and HCl^+ abstracts another H-atom from H_2 to form H_2Cl^+; this satisfies all the valencies of the Cl^+. The H_2Cl^+ ion can add no more hydrogen atoms, so the next most likely step is to recombine with an electron, forming hydrogen chloride, HCl:

$$H_2Cl^+ + e \rightarrow HCl + H$$

In general, however, the cold chemistry initiated by starlight acting alone is rather limited. This is because—of the most important elements, oxygen, carbon, nitrogen, and sulfur—only carbon (as C^+) reacts with molecular hydrogen—and that reaction is a rather slow radiative association. However, when cosmic rays also contribute, then new chemical pathways will open (see Section **3.3**).

3.2.2 X-Rays

Massive stars emit X-rays, and so do active galaxies (see Figure **3.1**). Therefore, it is important to understand whether X-rays can influence chemistry. X-rays are much more energetic than ultraviolet radiation, and are capable of ionizing all atoms and molecules. Near to a powerful source, X-rays will ionise every atom and molecule and suppress all chemistry. But 'hard' X-rays (with wavelengths less than about a nanometre) interact only weakly with matter, and so can cause a rather low level of ionisation but extended over very large volumes of interstellar space.

The chemistry initiated by this ionization is rather different from that initiated by starlight because every atom and molecule is

Figure 3.1 A false-colour image of the remnant of a supernova explosion (now called RCW 86) observed by Chinese astronomers in 185 AD. The image combines X-ray emission from gas heated to very high temperatures by the supernova shock wave, represented here as blue and green, with infrared emission from the much cooler dust, represented here as yellow and red. RCW 86 is in the plane of the Milky Way, about 2700 parsecs from the Sun and is clearly a powerful source of X-rays. (Credit: X-ray: ESA/XMM-Newton, NASA/CXC/SAO/Chandra/IR: NASA Wide-Field Infrared Survey Explorer, NASA JPL-Caltech, Spitzer Space Telescope.)

capable of being ionised by X-rays. In particular, ionisation of H_2 creates the molecular hydrogen ion, H_2^+, which is rapidly converted to H_3^+. As described in Section **3.1.5**, above, this ion readily donates a proton to other species, and initiates a rich chemistry.

Thus, for example, there are efficient routes to carbon monoxide, of which this is one:

$$C + H_3^+ \rightarrow CH^+ + H_2$$

$$CH^+ + H_2 \rightarrow CH_2^+ + H$$

$$CH_2^+ + H_2 \rightarrow CH_3^+ + H$$

$$CH_3^+ + O \rightarrow HCO^+ + H_2$$

$$HCO^+ + e \rightarrow CO + H$$

and forms another detected and important species, the formyl radical ion, HCO^+, in passing. The major difference from starlight-driven chemistry is that the X-rays (unlike ultraviolet radiation) are largely unaffected by the dust, so that X-rays can create ions deep inside dark clouds from which starlight is excluded. Thus, the important species C^+, C, and CO can coexist in regions affected by X-rays, while in regions affected by starlight these species are in fairly distinct zones, with C^+ ions in regions most heavily irradiated by starlight while C and CO predominate in regions more heavily shielded from starlight.

3.3 COSMIC RAYS

In Section **3.1.5** we described how the ionisation of molecular hydrogen by cosmic rays leads to the formation of protonated hydrogen, H_3^+. This is a key ion in interstellar chemistry, because it donates a proton very readily to many other species, creating new ions that lead to new species. We described how water and hydroxyl could be formed with the initial step being the donation of a proton from H_3^+ to an oxygen atom.

Similarly, carbon atoms can accept a proton from H_3^+ to form hydrocarbon and their ions, and the beginnings of chemical complexity:

$$C + H_3^+ \rightarrow CH^+ + H_2$$

followed by hydrogen abstraction reactions with H_2 molecules:

$$CH^+ \rightarrow CH_2^+ \rightarrow CH_3^+ \rightarrow CH_5^+$$

The formation of CH_3^+ is rapid, but the last stage is a slow radiative association. The end point is CH_5^+ because the valency of the C^+ ion is then fully satisfied. Finally, since CH_5^+ can no longer react with the abundant H_2, it dissociatively recombines with electrons to form methane:

$$CH_5^+ + e \rightarrow CH_4 + H$$

The beginnings of complexity can be envisaged through reactions such as:

$$CH_3^+ + C \rightarrow C_2H^+ + H_2$$

and the C_2H^+ ion leads to a series of molecules containing two carbons, such as diatomic carbon (C_2), the ethynyl radical (C_2H), and acetylene (C_2H_2). These species are important in building up even greater chemical complexity, especially in carbon chains. For example, the following reaction forms the linear molecule diacetylene (H_2C_4):

$$C_2H + C_2H_2 \rightarrow HC_2 - C_2H + H$$

and longer carbon chains are formed in similar reactions.

Nitrogen, however, continues to present some difficulty; what are the reactions initiating the chemistry of nitrogen-bearing species? The reaction of H_3^+ with N is endothermic and does not occur in cold interstellar clouds. As an alternative route, cosmic rays may create N^+ ions by ionising molecular nitrogen (N_2) or other N-containing molecules. Reactions with He^+ also produce N^+ ions, which can react with H_2 molecules to form NH^+ ions, which could provide an entry to the chemistry. However, even this route is inhibited at the low temperatures of dark clouds (though it is a possible route in the warmer diffuse clouds).

Therefore, it appears more likely that nitrogen chemistry in dark clouds, as in diffuse clouds, is initiated (at least, in part) by reaction of N-atoms with radicals such as CH or OH to form CN or NO, and C may exchange with NH to form the cyanide radical.

The insertion of nitrogen atoms into partially unsaturated hydrocarbon chains is another way to form the strong cyanide bond. The CN radical has a large dipole moment, making carbon chain molecules containing CN radicals into powerful radiators in the microwave spectrum. For example, the following reaction forms cyanoacetylene (HC_3N), a widely detected carbon chain species in interstellar clouds:

$$N + C_3H_3 \rightarrow HC_3N + H_2$$

Cyanoacetylene is a member of a remarkable series of polycyanoacetylenes in the interstellar and circumstellar media, of which the largest detected member is $HC_{11}N$. These are linear carbon-chain molecules, with a cyanide radical at one end.

The most important cyanide-bearing species in the interstellar gas are hydrogen cyanide (HCN) and hydrogen isocyanide (HNC). Hydrogen cyanide can be formed in a wide variety of ways, including atom exchanges with hydrides, such as the following:

$$N + CH_2 \rightarrow HCN + H$$

$$N + CH_3 \rightarrow HCN + H_2$$

$$C + NH_2 \rightarrow HCN + H$$

and reactions of HCN with H_3^+ form H_2CN^+ ions which when dissociatively recombining with electrons create CN, HCN, and hydrogen isocyanide, HNC. Thus, it appears that the formation of nitrogen-containing molecules in interstellar clouds depends on the existence of precursor molecules whose chemistry is driven by cosmic ray ionisation. These nitrogen-containing species are sometimes called 'late-time' species, while the hydrocarbons formed earlier are called 'early-time'. In principle, therefore, chemistry provides a 'clock' by which one can assess the evolutionary status of an interstellar cloud as it develops from a low-density diffuse cloud through a dark cloud and into a dense state in which star-formation may occur. The timescales for chemistry and for dynamical change are probably similar, at least in clouds in the Milky Way galaxy.

Sulfur is another species incapable of reacting with H_3^+ at low temperatures; so sulfur-bearing species are created in similar ways to those forming nitrogen molecules. Sulfur atoms undergo exchange reactions with OH to form sulfur monoxide (SO) and reactions of SO with OH create sulfur dioxide (SO_2):

$$S + OH \rightarrow SO + H$$

$$SO + OH \rightarrow SO_2 + H$$

Both SO and SO_2 are important detected interstellar species, found in denser regions of the interstellar medium. Reactions of S-atoms with CH form carbon monosulfide (CS), with CH_2 form thioformyl radical (HCS), and with the methyl radical (CH_3) form thioformaldehyde (H_2CS).

Further reactions make more complex species, such as carbonyl sulfide (OCS), which may form in reactions:

$$CO + SH \rightarrow OCS + H$$

$$O + HCS \rightarrow OCS + H$$

Clearly, sulfur-bearing species should be regarded as 'late-time' species.

Cosmic rays also create other ions that are important in the chemistry. As well as ionising hydrogen molecules, they can also ionise hydrogen atoms directly to form protons, *i.e.* H^+ ions, and a minor channel of ionisation of H_2 molecules also leads to H^+ formation. There is a curiosity associated with hydrogen and oxygen atoms: their ionisation potentials are—accidentally— almost exactly equal; that of oxygen is very slightly larger, the difference in energy—converted to a temperature—being about 226 K. This means that at temperatures much below this value the reaction between protons and oxygen atoms is suppressed, while at temperatures approaching and above this value oxygen ions can be created:

$$H^+ + O \rightarrow H + O^+$$

So protons created by cosmic rays can form O^+ ions, which do react directly and efficiently with molecular hydrogen:

$$O^+ + H_2 \rightarrow OH^+ + H$$

and the OH^+ ions may then go on to form water or hydroxyl as previously described in Section **3.1.5**. This route is not important in dark clouds where the temperature is too far below the critical value of 226 K. But in diffuse clouds, with temperatures of around 100 K, a sufficient number of collisions between H^+ and O may have enough energy to initiate this important route to H_2O and OH.

Helium atoms are present in interstellar gas, at about 10% by number of hydrogen atoms. Helium has a high ionisation potential (24.6 eV), and it is far beyond the ability of starlight to ionise helium. But cosmic rays (and X-rays) easily create He^+ ions, which have a dramatic effect on interstellar chemistry. For, with this high ionisation potential, He^+ is capable of ionising any other species and almost certainly destroying it (see Section **2.3.3**). The reaction:

$$He^+ + CO \rightarrow He + C^+ + O$$

is an example of dissociative charge transfer which leaves the CO^+ molecule in an unstable condition, so it dissociates. In fact, He^+ ions are capable of ripping apart in this fashion any known interstellar molecules. Helium ions are the sharks of interstellar chemistry.

3.4 DUST GRAINS

Although gas phase reactions contribute enormously to the richness of astrochemistry, reactions on the surfaces of dust grains also have a very important role. These reactions lead to the formation of simple molecules, and the accumulation of simple molecules as ices permits the production of the most complex of astronomical molecules.

3.4.1 Molecular Hydrogen

As we saw in Chapter **2**, the early speculation that reactions on dust grain surfaces between H-atoms could form hydrogen molecules has been supported by laboratory experiments and theoretical investigations. Clearly, dust grains supply molecular hydrogen to the interstellar medium, and are therefore ultimately responsible for driving a gas phase interstellar chemistry initiated by reactions with hydrogen molecules, as discussed in Section **3.1**. Evidently, dust grains are effective catalysts in H_2 production. What other chemical properties do they have?

3.4.2 Dust Grains as Catalysts

There is strong astronomical evidence (and an appropriate laboratory experiment, see Section **9.3.2**) supporting the view that in dark clouds oxygen atoms can be converted to water molecules on dust grain surfaces, with the product H_2O molecules being retained so that grains become coated with mantles of ice. These ice mantles are detected by absorption spectroscopy in the near infrared, through pure vibrational transitions of H_2O near 3 microns (rotation being suppressed in the solid), along lines of sight through darker interstellar clouds where dust provides considerable shielding to ultraviolet starlight. The ice that is formed also contains relatively large amounts of other species, notably carbon monoxide and carbon dioxide (at about the 20% level relative to H_2O by number), and lesser amounts of other species. These and other species are all detected in the ice through their pure vibrational transitions (see Figure **3.2** and Table **3.3**).

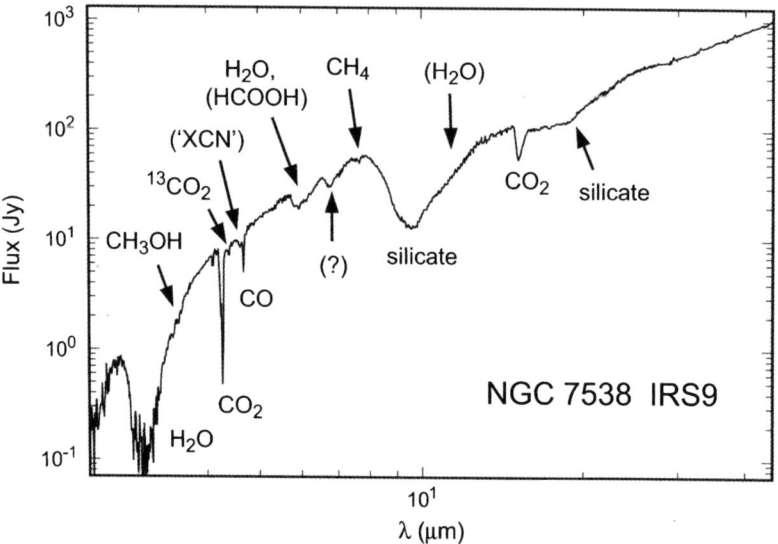

Figure 3.2 The 3–20 μm wavelength spectrum towards the infrared source NGC 7538 IRS9, using the Infrared Space Observatory satellite. Dust along the line of sight to IRS9 absorbs solid-state features corresponding to stretching and bending vibrational modes of various species in the composition of the dust. The 'silicate' features near 10 and 20 μm correspond to Si–O stretching and bending modes, respectively, in solid silicates. Reproduced with permission from D. C. B. Whittet *et al.* 1996, Astronomy & Astrophysics **315** L35 7. Copyright ESO.

Table 3.3 The composition of interstellar ices varies very significantly from one line-of-sight to another. Species near the top of the list are relatively abundant in the ices and securely identified, with CO and CO_2 typically around 20% of H_2O, by number. Species towards the bottom of the list are much lower in abundance. XCN signifies a CN-bearing species.

H_2O
CO
CO_2
CH_3OH
H_2CO
HCOOH
NH_3
CH_4
XCN
OCS

Neither water nor carbon dioxide have significant abundances in the gas phase in the interstellar clouds where the ice is detected. But the amount of ice present on the dust grains may be quite large, containing a significant fraction of the available oxygen. This large amount of H_2O ice could not be accumulated from the gas phase within the lifetime of the cloud. On the other hand, the amount of water ice in a cloud is consistent with the formation of H_2O molecules by the efficient hydrogenation of O-atoms incident on the dust grain surface.

Similarly, the amount of carbon dioxide in the ice far exceeds the amount that could be accumulated from the gas phase within the lifetimes of the cloud, as CO_2 is a very rare species in the gas phase. We infer that carbon dioxide must also be created *in situ*, through surface reactions involving carbon monoxide, such as:

$$mCO + mOH \rightarrow mCO_2 + mh$$

where the prefix '*m*' indicates a mantle species. The origins of the species mH_2O and mCO_2 are therefore to be regarded as distinct from that of mCO; carbon monoxide is abundant in the gas and there is no difficulty in accounting for the amount of carbon monoxide that is in the ices, through collisions with grains and sticking to their surfaces.

The ices also contain methane and ammonia in amounts that are larger than can be deposited from the gas phase in an appropriate timescale. Again, one is tempted to infer that methane and ammonia are created *in situ* and retained on the grains, contributing to the ice composition. However, while the formation and retention of water, carbon dioxide, methane and ammonia on interstellar dust grains is strongly indicated by astronomical data, there is limited laboratory evidence under appropriate conditions supporting these presumed conversions.

There are, however, some experiments (see Section **9.3.2**) that address the detection of two other solid-state species in interstellar ices: formaldehyde (H_2CO) and methanol (CH_3OH). These species are present at the level of a few percent relative to the number of H_2O molecules. Water in the ice takes on the order of ten percent of the available oxygen, or about 10^{-4} relative to hydrogen. Thus, formaldehyde and methanol in the ice have an abundance that is a few times 10^{-6} relative to hydrogen. This exceeds the amount of formaldehyde and methanol in the gas phase. Hence, it seems that these two species are supplemented by reactions in or on the ices.

Several laboratory experiments now support the idea that carbon monoxide can be hydrogenated in stages, producing both formaldehyde and methanol in sequence:

$$mCO \rightarrow (mHCO + mCOH) \rightarrow mHCOH \rightarrow mCH_2OH \rightarrow mCH_3OH$$

The chemical composition of ices along quiescent paths through the interstellar medium can therefore be broadly understood in terms of simple accretion and hydrogenation of atoms and molecules, and the oxidation of mCO to form mCO_2.

3.4.3 Chemical Processing of Interstellar Ices

Ices may remain on dust grains in dark clouds for some millions of years. Towards the end of that period of quiescent evolution, the cloud may be affected by the formation of one or more stars within it. The material near the newly formed stars will be heated by starlight, the gas and grains will warm up and the ices evaporate. Does the ice remain unchanged during this period? What processes, if any, can have occurred in the ice? Is some kind of solid-state chemistry possible in ices as their temperature rises?

It seems likely that some of the most dramatic of all interstellar chemistries occur in these evolutionary situations in interstellar clouds. The relatively simple ices that we have described contain almost all the important elements O, C, N, and S in the regions where the ices are detected; very little is left in the gas phase. So if chemistry is to occur in such regions—and, evidently, it does—the chemistry must occur in the material of the ices. Therefore, the ices must be the feedstock for a complex chemistry that gives rise to the molecular complexity that we described in Chapter 1, and is found by astronomers in star-forming regions, including that at the centre of the Milky Way galaxy.

Evidently, the grains—acting as a substrate on which ices can form—enable a remarkable conversion to occur. Somehow, simple species in the originally deposited ice (H_2O, CO, CH_3OH, H_2CO, NH_3, CH_4, *etc.*) must generate the relatively complex molecular species observed, such as methyl formate [$HCOOCH_3$], dimethyl ether [CH_3OCH_3], acetone [CH_3COCH_3], ethylene glycol [$(CH_2OH)_2$], and glycolaldehyde [$CH_2(OH)CHO$]. We will discuss in Sections **4.3.1** and **4.3.2**, how this conversion may occur.

3.4.4 Surface Reactions in Diffuse Clouds?

The presence of ices in dark interstellar clouds appears to be strong evidence in support of a variety of reactions occurring on the surfaces of dust grains. However, these ices are found only inside interstellar clouds where the extinction due to dust is high, probably with an optical depth in the ultraviolet of about 10. Does this mean that surface reactions other than H_2 formation occur only in these dark regions? Do such reactions also occur in, for example, diffuse clouds, through which starlight penetrates fairly easily?

In the case of H_2 formation on dust grains, the surface reactions are certainly taking place in diffuse clouds (and dark clouds, too, of course). Yet, the absence along lines of sight with low dust extinction of absorptions near 3 microns, the signature of pure vibrational transitions in H_2O (see Figure **3.2**), suggests that rather little—if any—ice accumulates on dust grains in diffuse clouds (see Figure **3.3**). So it has been conventional to ignore any possible

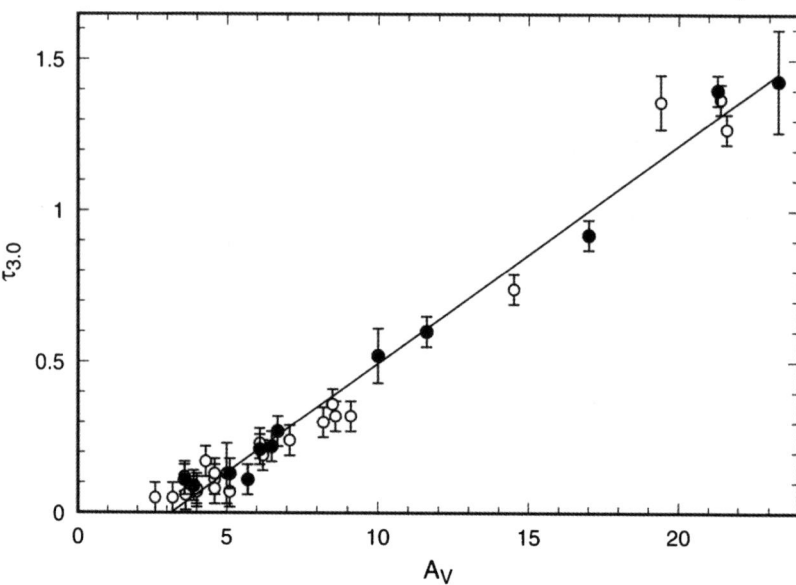

Figure 3.3 A plot of the optical depth of the H$_2$O-ice absorption on various lines of sight through the Taurus molecular cloud, as a function of visual extinction, A$_V$, along the path. The data shown are a compilation of results from various observations. Evidently, H$_2$O-ice is unlikely to be present along lines of sight in Taurus for which the visual extinction is less than about 3 magnitudes. Reproduced with permission from D. C. B. Whittet *et al.* 2001, Astrophysical Journal **547** 872. Copyright American Astronomical Society.

contribution to interstellar chemistry in diffuse clouds, apart, of course, from molecular hydrogen.

However, this conventional view may be challenged by the recent detection (by the Herschel Space Observatory) in diffuse clouds of the hydrides of nitrogen: NH, NH$_2$ and NH$_3$. Given the difficulty in making ammonia (NH$_3$) in the gas phase (see Section 3.3), the presence of these species suggests that surface reactions may contribute ammonia to the gas phase, where photodissociation by starlight produces the amino radical (NH$_2$) and the monohydride (NH). In fact, nitrogen monohydride had been detected some ten years earlier, but because of the difficulty of understanding nitrogen chemistry the detection was regarded as inconclusive evidence for surface reactions.

If nitrogen is hydrogenated at surfaces of dust grains in diffuse clouds, do oxygen atoms also undergo a similar conversion to H$_2$O molecules? What about carbon? It exists in diffuse clouds in the form of carbon ions, C^+; does this state affect any possible hydrogenation of carbon, or is the carbon simply incorporated into carbonaceous solids in dust grains? At present, there are no experimental results carried out under appropriate physical conditions to help provide answers to these questions.

However, this picture raises an interesting chemical problem. Why is it that molecules such as H_2O, NH_3, and CH_4, when made on dust grains inside dark clouds, are retained as ices, whereas—since the ices are not detected in diffuse clouds—the molecules must be (mostly) ejected into the gas phase in diffuse clouds? There is no clear understanding of this difference between diffuse and dark clouds. It is possible that photodesorption by starlight in diffuse clouds may release the molecules. Or it may be that dust grain surfaces have different structures (and hence different binding energies for hydrides) in the two types of cloud, with those in the dark clouds being more favourably inclined to retain the products of surface reactions. Alternatively, hydrogen bonding between hydride molecules may simply mean that the greater surface coverage expected in dark clouds (where the gas density is much greater than in diffuse clouds) encourages ices to build up, while in diffuse clouds a critical surface coverage is not achieved.

3.5 DYNAMICAL HEATING

In the mainly neutral interstellar gas where molecules are found, the gas temperature (determined from molecular emissions, see Chapter **8**) is generally rather low. In diffuse clouds the temperature is typically around 100 K, while in dark clouds (which contain most of the interstellar mass and are opaque to starlight) the gas temperature is about 10 K. Of course, there are regions of interstellar space close to hot stars (see Figure **1.4**) where the temperatures are high—about 10 000 K—and some extensive and very low density regions heated by supernova shocks that are at very high temperatures—around 1 000 000 K, but these are ionised regions where molecules do not exist.

We saw in Section **3.1** that the low temperatures of mainly neutral regions mean that direct exchange reactions between, say, O-atoms and H_2 molecules to form OH radicals are severely inhibited. However, if regions of higher temperature exist in these mainly neutral regions, then perhaps these and other exchange reactions may be important. After all, H_2 molecules and O-atoms are two of the most abundant interstellar species, so if they can react, their contribution will be significant.

There are two main ways that we will describe in which normally cold interstellar gas can be heated without using intense radiation fields of starlight or of cosmic ray particles. These ways of heating depend on gas dynamics. Interstellar clouds are always observed to be in motion (with velocities relative to one another on the order of ten kilometres per second), so that when different parcels of gas meet with relative velocities exceeding the local sound speed (typically, less than about 1 km per second), a shock wave results. The shock turns some of the energy of motion into thermal energy in the shock, and temperatures in a small region of gas can be raised abruptly to thousands of K, or even greater.

The second heating mechanism is turbulent mixing that occurs when an outflow of hot ionised and molecule-free gas from a star rushes past a cold mainly neutral cloud. The interaction between the two gases is not a smooth laminar flow, but a highly turbulent layer. Hot ionised gas can be mixed with cold neutral gas in varying proportions across the turbulent layer, and the insertion of heat and ionisation from one gas into the other can affect the chemistry, with distinctive results.

3.5.1 Shocks

Cold non-magnetic molecular gas clouds colliding with relative velocities of 10 km s^{-1} suffer a shock in the collision zone where the temperature will rise abruptly to several thousand K. This high temperature leads to collisions between molecules that are energetic enough to populate excited vibration–rotation levels in molecules such as carbon monoxide, water, and hydroxyl. These levels then radiate quickly, and so the gas cools, typically on a timescale of a hundred or a thousand years. If the dynamics of the shock is magnetically moderated, the temperature rise is less abrupt, reaches a somewhat lower but still elevated temperature of about a thousand K. However, this elevated temperature persists for a longer time during which kinetic energy is converted to heat. In either the magnetic or non-magnetic cases, it is in this relatively short period of elevated temperature that chemical pathways inhibited at lower temperatures may open. Of course, if the shock speed is too great then the resulting temperature may be so high that molecules are collisionally destroyed very quickly, and all of chemistry is suppressed. A crucial factor concerns molecular hydrogen. If it is collisionally destroyed in the hot post-shock gas the resulting hot H-atoms are capable of breaking down any interstellar molecule. Even a strongly bound molecule such as carbon monoxide is reduced by hot hydrogen in reactions that would at low temperatures be endothermic; for example:

$$H^* + CO \rightarrow C + OH$$

$$H^* + OH \rightarrow H_2 + H$$

$$H^* + H_2 \rightarrow 3H$$

where the star (*) denotes that the H-atom carries a large amount of energy. For a non-magnetic case shock speeds up to about 25 km s^{-1} permit a shock chemistry to occur; above this value the H_2 dissociates and the hot H-atoms destroy all other molecules. For a magnetic shock, the critical velocity is about 45 km s^{-1}.

The chemical effects of a shock with velocity less than the critical value are dramatic. The reactions of atoms, molecules and ions with molecular hydrogen that are endothermic at low temperature become accessible. Water is formed directly and rapidly by:

$$O + H_2 \rightarrow OH + H$$

$$OH + H_2 \rightarrow H_2O + H$$

in reactions that become about one percent efficient; H_2O molecules are expected to take up almost all the oxygen not already bound in CO.

Carbon atoms may undergo similar reactions, but these have even larger barriers, so that they do not contribute much to carbon chemistry. Carbon ions, however, react very efficiently with H_2 at temperatures above 1000 K and the reaction:

$$C^+ + H_2 \rightarrow CH^+ + H$$

in warm regions is undoubtedly the source of interstellar CH^+, detected in diffuse clouds. In regions rich in H_2, reactions in shocked regions of CH^+ with H_2 are very fast and ultimately contribute to the formation of simple hydrocarbons such as CH, CH_2, and CH_3 (see Section **3.3**).

Nitrogen atoms in modest shocks can react with H_2 to make NH and NH_2 fairly efficiently, but the step to NH_3 is slow.

Sulfur is an interesting case. In diffuse clouds, sulfur is present as S^+ ions, which react successively with hydrogen molecules fairly efficiently in shocks, to form H_3S^+. Dissociative recombination of this ion with electrons produces sulfur monohydride (SH) and hydrogen sulfide (H_2S), and reactions of these species with O-atoms makes SO and SO_2. In dark clouds, the sulfur is largely in the form of neutral atoms. Abstractions of hydrogen from H_2 in shocked regions again create HS and H_2S which are the precursors of SO, SO_2, and other sulfur-bearing species. As we have seen in Section **3.3**, entry to sulfur chemistry in cool regions is generally difficult. Therefore, an abundance of sulfur-bearing molecules may be regarded as a signature of shocks.

The effect of the passage of a shock on the chemistry of a dark cloud (for a shock speed within the limits discussed above) is to convert the cloud into a new, transient and unstable chemical state. In this new state, nearly all free atoms and ions are hydrogenated in reactions with H_2 molecules, and some conversion of these hydrides will occur to species containing O, C, N and S atoms. After the shock has passed and the physical conditions have reverted to those pertaining pre-shock, the chemistry of the cloud will also revert to it pre-shock chemical condition. However, the return of the cloud to its original chemical state takes much longer than the period (around a thousand years) for the transition into the shocked state. The reversion to the normal chemical condition in a dark cloud is related to the rate of the ion-molecule chemistry driven by cosmic ray ionisation. For physical conditions appropriate to the Milky Way galaxy, this may take up to a hundred thousand years. Thus, the shocked chemistry may persist for a significant part of the evolution of a cloud.

3.5.2 Interfaces

A wide range of physical conditions exists across an interface. At one extreme, there may be a hot ionised stellar wind, while at the other there may be a cold dark cloud containing dust grains coated with icy mantles. Within the interface, turbulence mixes these different regimes together, supplying heat and ionisation to the cold gas and molecular-rich material to the hot gas. The result is a non-standard situation for chemistry; although it is hotter than normal, it is not the same as shock chemistry because of the enhanced ionisation, and it is not like a diffuse cloud (where ionisation is relatively high) because of the absence of a significant ultraviolet radiation field and of the presence of molecular material from a dark cloud. So, it is a special situation and, therefore, not a surprise that these conditions enhance a chemistry that produces a range of molecular species that we can regard as a signature of interfaces.

The first molecule whose emission was attributed to arising in an interface was the formyl radical ion, HCO^+. The emission was detected towards young low-mass stars and the emission region had an unusual morphology—a so-called 'butterfly' shape. Also, the inferred abundance of the HCO^+ was unusually high. It was realised that the shape corresponded with what might be observed if a stellar wind was blowing a cavity in a dense core surrounding the star and the emission was occurring in the boundaries of the cavity. 'Butterfly' morphologies arising from interface zone around a cavity swept by a stellar outflow may also be seen in other emission lines, see Figure **3.4**.

The high abundance was initially not understood; as we have seen (*e.g.* in Section **2.3.2**), HCO^+ is conventionally formed by the donation of a proton from H_3^+ to carbon monoxide:

$$H_3^+ + CO \rightarrow HCO^+ + H_2$$

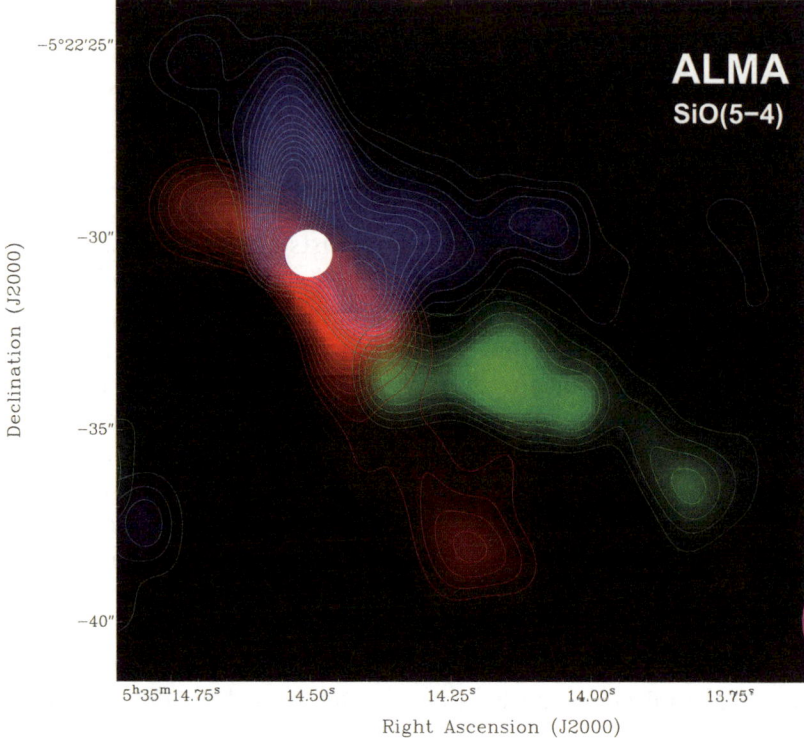

Figure 3.4 An image of 'butterfly' morphology in SiO emission in the 5–4 rotational transition, with data from the Atacama Large Millimeter Array (ALMA) of Orion Source 1. The red and blue contours are from material moving away and towards us. The SiO is probably released when silicate dust is destroyed in the interface between the outflow from a very young star and the cloud in which the star was formed. The green contours are of material in the same line of sight, but not necessarily associated with Source 1. Reproduced with permission from L. Zapata *et al.* 2012. Astrophysical Journal **754** L17. Copyright American Astronomical Society.

and to make more HCO^+ than normal one needs more H_3^+, and therefore a higher cosmic ray ionisation rate (see Section **3.3**). But more cosmic ray ionisation means more electrons and consequently a faster dissociative recombination of HCO^+:

$$HCO^+ + e \rightarrow H + CO$$

As a consequence, the HCO^+ abundance is not very sensitive to changes in the cosmic ray flux. But in an interface we have a new source of HCO⁺ that does not normally play much of a role in dark clouds:

$$C^+ + H_2O \rightarrow HCO^+ + H$$

where the C^+ comes from the ionised gas and the H_2O comes from the ices that are released from mantles in the warmer interface zone. This additional formation route enables higher HCO⁺ abundances to be established in the interface, and so the observations support the view that mixing is occurring in the interface. In fact, HCO⁺ is not the only species that should be enhanced in such an interface. Detailed models show that interfaces near low-mass stars should be rich not only in HCO⁺ but also in hydrogen sulfide (H_2S), carbon monosulfide (CS), thioformaldehyde (H_2CS), sulfur monoxide (SO), and sulfur dioxide (SO_2). These sulfur-bearing species are enhanced because the special physical conditions in the interface enable the barriers to sulfur chemistry in normal clouds (see Sections **3.2** and **3.3**) to be more readily overcome. Methanol (CH_3OH) is also likely to be enhanced in interfaces near low-mass stars, because of the surface reactions that have hydrogenated CO in the ice (see Section **3.4.2**) and the subsequent injection of ices into the warm interface.

Chemical enhancements of unusual morphology have also been found near to massive stars, where the physical conditions may differ from those near to low-mass stars. We expect a much higher gas density near high-mass stars, and also an intense ultraviolet radiation field (not present in the low-mass case). Observations of one particular high-mass star-forming region show that an extended morphology exists in thioformaldehyde (H_2CS), carbonyl sulfide (OCS), methanol (CH_3OH) and deuterated water (HDO), but is absent in some other species (H_2S, SO, SO_2, and CS) that are certainly present but do not show the extended morphology. Chemical models support the view that this behaviour should arise in a UV-irradiated high-density interface.

3.6 CONCLUSION

We have shown that there are four important routes to interstellar chemistry: through the effects of starlight, cosmic rays, dust grains, and dynamics. These routes, initiated by these four 'drivers', generate characteristic chemical responses. Usually, of course, more than one driver may be acting at the same time. For example, in diffuse clouds starlight, cosmic rays and dust grains act all the time, while from time to time the diffuse gas will be influenced by a shock wave. One of the fascinating tasks for the astrochemist is to use the information present in molecular spectra to determine the effectiveness of each of the different drivers acting in different regions. Here, it is useful to note that special physical conditions may, through combinations of drivers, give rise to special signature chemistries (as we have seen in interfaces, for example, in Section **3.5**).

Of course, in different locations the physical conditions may be very different. In Chapter **4** we will discuss chemistry in the Milky Way galaxy, *i.e.* our galaxy and the home that has nurtured the Sun, its Solar System, and our planet Earth. Other galaxies may have rather different physical conditions, such as very intense radiation fields, or different elemental abundances. We discuss chemistry in such galaxies in Chapter **6**. But wherever chemistry is occurring in interstellar space one or more of the drivers that we have discussed in this chapter is driving it.

The discussion in the chapter has been presented in terms of interstellar chemistry. The ideas presented also apply to circumstellar chemistry, though in these regions the gas densities may be much higher than in interstellar clouds, and the physical conditions may also be very different. The atmospheres of some stars may be cool enough for molecules to exist in abundance in them. In fact, the densities in the atmospheres, and the temperatures, are likely to be high enough for three-body reactions to play a part. For example, the reaction of three hydrogen atoms to form a hydrogen molecule and a hydrogen atom (where the atoms carries away excess energy that allows the molecule to be stable):

$$H + H + H \rightarrow H_2 + H$$

begins to be important for number densities of H atoms of about 10^{11} per cm^3, if the temperature is on the order of 1000 K or so. Three-body reactions occur with other partners, too. In warm stellar atmospheres, exchange reactions between molecules continue until the atoms are bound together in the most energetically favourable arrangements, consistent with the physical conditions. For cool carbon-rich stars, the most abundant species in the atmospheres include CO, C_2H_2, CH_4, NH_3, HCN, and N_2. In an oxygen-rich star, the most abundant species include H_2O, CO_2, and NO. These molecules are of course accompanied by and greatly outnumbered by molecular hydrogen, in each case.

FURTHER READING

T. W. Hartquist and D. A. Williams, *The Chemically Controlled Cosmos*, Cambridge University Press, 1995.

T. W. Hartquist and D. A. Williams, eds, *The Molecular Astrophysics of Stars and Galaxies*, Oxford Science Publications, Clarendon Press, 1998.

CHAPTER 4
Molecules in the Milky Way Galaxy

4.1 DIFFUSE INTERSTELLAR CLOUDS

Diffuse clouds were the first astronomical locations in which interstellar molecules were identified. Molecular spectra, later attributed to radicals methylidyne (CH) and cyanogen (CN) and the methylidyne ion (CH^+), were detected in 1937 in absorption in the optical spectra of distant bright stars. At the time, this was a surprise because interstellar conditions were thought to be too hostile for molecules to exist in detectable amounts. Indeed, the subject of astrochemistry began with the attempts to understand the origin of molecules in diffuse clouds. As the chemical networks in diffuse clouds are rather simpler than those of, say, star-forming regions, diffuse clouds continue to be much studied and they provide crucial tests of important astrochemical ideas.

Following these early detections, now—more than seventy years later—we know that a wide variety of mainly diatomic and triatomic species exist in diffuse clouds (see Table **4.1**). New discoveries are still being made; most recently, the Herschel Space Observatory identified the ions OH^+, H_2O^+ and H_3O^+ and—as mentioned in Section **3.4.4**—the neutrals NH, NH_2 and NH_3 in such regions.

We have summarised the physical conditions in diffuse clouds in Table **1.2**. We can see from that information that diffuse clouds

---

The Cosmic-Chemical Bond
DA Williams and TW Hartquist
© DA Williams and TW Hartquist 2013
Published by the Royal Society of Chemistry, www.rsc.org

Table 4.1 Some of the molecular species that have been detected in diffuse interstellar clouds by spectral line absorptions in the visible and ultraviolet, or by emissions in the millimetre and submillimetre-wave regions of the electromagnetic spectrum.

H_2, HD, H_3^+
OH^+, OH, H_2O^+, H_2O, H_3O^+
CH^+, CH
NH, NH_2, NH_3
HCl, H_2Cl^+
HF
CO,
C_2, C_2H
CN, N_2
HCO^+, HOC^+
H_2CO
C_3, c-C_3H_2

are cold, with temperatures around 100 K, and the gas is very tenuous indeed, typically with about 100 H-atoms available per cm^3 (recall that the atmosphere we breathe has about 2.7×10^{19} molecules per cm^3). The gas in interstellar diffuse clouds is a mixture of H-atoms and H_2 molecules, with a range of H/H_2 values, some clouds being mostly atomic and others mostly molecular. There is, of course, a minor component of elements such as oxygen, carbon, nitrogen, sulfur, *etc.* (see Section **1.2**, Table **1.1**), some of these elements being incorporated into dust grains. Only a small amount of extinction of visible starlight due to dust is present in these objects. Typically, diffuse clouds have an optical depth in the visible of less than or about unity, meaning that the visible starlight intensity is reduced inside the cloud only by a maximum factor of up to three or so. This last parameter is an important one, for it implies that starlight streams rather freely through such clouds. The mean intensity of unshielded starlight in the interstellar medium is strong enough that a fairly typical photodissociation such as:

$$CH + starlight \rightarrow C + H$$

gives a lifetime for the CH molecule of only a few hundred years in unshielded regions, and perhaps ten times longer in the interiors of diffuse clouds. Given that interstellar gas densities are so low, we need to find mechanisms for forming the detected molecules that are efficient. We will look in some detail at the chemistry of diffuse clouds, not because these clouds are more important than other types of region in an astronomical context, but because the discussion we shall make illustrates many of the issues that arise in the chemistry of other interstellar regions.

We will assume—as indicated by the observations and informed by our discussion in Chapter **2**—that hydrogen molecules formed in surface reactions on dust grains take up a significant amount of the available hydrogen, the rest being present as H-atoms. The trace elements—oxygen, carbon, nitrogen, *etc.*—are present as atoms. So, starting from our discussion in Chapter **3**, the problem is to find efficient pathways in such a gas to make the molecules indicated in Table **4.1**, where the abundances of the trace elements relative to hydrogen are assumed to be the values we find in the Sun, as given in Table **1.1**. It is important to note that these pathways must be effective at the low temperatures of interstellar clouds. As pointed out in Section **1.3**, many direct exchange reactions between atoms and molecular hydrogen are strongly suppressed at the low temperatures of interstellar clouds.

4.1.1 Gas Phase Chemistry in Diffuse Clouds

We will look first at two sources of energy that may help to promote chemistry in diffuse clouds: these are starlight and cosmic rays. Both of these sources can induce ionisation, and the ions produced may help to initiate chemistry, even at low temperatures. Grains surface reactions are obviously invoked for H_2 formation.

As discussed in Section **3.2.1**, atoms with ionisation potentials less than that of hydrogen (*i.e.* less than 13.6 eV) can be ionised by ultraviolet radiation in the interstellar radiation field. This means that the following relatively abundant elements are normally ionised in diffuse clouds: carbon (C), sulfur (S), silicon (Si), iron (Fe), magnesium (Mg), calcium (Ca), sodium (Na), potassium (K), chlorine (Cl), and calcium ions (Ca^+) are normally ionised a second time (Ca^{++}). However, the most important elements, oxygen and nitrogen are mainly neutral in diffuse clouds. Most of the metals are probably locked up in dust grains, and not available for gas-phase chemistry.

For some of these ions, an exchange reaction with H_2 occurs on nearly every collision. For example, the reaction of Cl^+ with H_2 initiates the formation of hydrogen chloride (HCl), a detected species in diffuse clouds.

In fact, even reactions between ions and H_2 are often suppressed. For example, there is a barrier to the reaction of sulfur ions (S^+) with H_2 molecules equivalent to almost 10 000 K, so those reactions are suppressed at the low temperatures of diffuse clouds. Similarly, for the most abundant atomic ion, C^+, there is a barrier to reaction with H_2 of 4640 K, meaning that such reactions are strongly suppressed at temperatures of about 100 K; so we cannot initiate carbon chemistry in diffuse clouds in that way.

In fact, the most effective entry route to carbon chemistry appears to be the rather inefficient radiative association of C^+ ions with H_2 to form the detected species methylidyne (CH), (see Section **3.1.3**). Then, neutral exchange reactions of CH with O and N atoms are efficient and can form CO and CN, both of which are detected species in diffuse clouds. Sulfur is also ionised and similar radiative associations of S^+ with H_2 lead to the formation of sulfur monohydride (SH) and hydrogen sulfide (H_2S).

However, similar routes stimulated by starlight to hydrogenate oxygen and nitrogen atoms, do not exist, because O^+ and N^+ are not available. For other ways of entering the chemistry, we need to look at the effects of cosmic rays on interstellar gas. Without cosmic rays, the chemistry of diffuse clouds predicted by the chemical models would be rather sparse when compared with the observational detections (Table **4.1**).

As described in more detail in Chapter **3**, the important ions created by cosmic rays are H^+, H_2^+, and He^+. The first two provide entry routes into oxygen chemistry. The third, He^+, is capable of destroying any molecule.

In diffuse clouds containing some atomic hydrogen, cosmic rays create H^+ ions which, in gas with a temperature above about 100 K, can exchange electrons with oxygen atoms to form O^+ ions which take part in exchange reactions with hydrogen molecules and form OH^+ (see Section **3.3**). This product then reacts in two further steps with molecular hydrogen to form H_2O^+ and H_3O^+ (which are detected species) At this point, the O^+ ion can add no more H atoms, as H_3O^+ has a complete electronic shell. In diffuse clouds the most likely reaction, occurring on almost every collision, is dissociative recombination with electrons and the product H_2O is also detected in diffuse clouds. The radiation field will dissociate H_2O in two stages, first to OH (a detected species in diffuse clouds) and then to O-atoms.

In diffuse clouds dominated by H_2 molecules, the cosmic rays mainly create H_2^+ ions (see Section **2.3.2**), which react promptly with another H_2 molecule to create H_3^+ ions. They readily donate a proton to many other species, including O-atoms, forming OH^+ ions, which then goes through the chemical stages described above. Both these routes provide through OH and H_2O an entry to the chemistry of oxygen-bearing molecules. For example, the routes involving OH:

$$C^+ + OH \rightarrow CO^+ + H$$

$$CO^+ + H_2 \rightarrow HCO^+ + H$$

$$HCO^+ + e \rightarrow H + CO$$

provide the detected species CO and HCO^+. Similar routes involving H_2O also occur.

The entry to the chemistry of nitrogen-bearing molecules is more problematic. Nitrogen atoms do not react with H_2, H^+, or H_3^+ at low temperatures, so the chemical routes described above that apply to oxygen do not work with nitrogen. However, N atoms do undergo exchange reactions at low temperatures with some partners, as discussed in Section **3.3**, to form such species as CN and NO. This is the main chemical entry route for nitrogen-bearing species in diffuse clouds, and similar reactions produce a variety of sulfur-bearing species. But we cannot make the detected species NH, NH_2, and NH_3 by these routes. As discussed in Section **3.4.4**, they may be formed in surface reactions. A simple calculation suggests that if ammonia (NH_3) is injected into the gas of diffuse clouds by surface reactions at an efficiency that is equivalent to the H_2 formation efficiency, and that the product NH_3 molecules are subjected to photodissociation by the interstellar radiation field, then the abundances of NH_3, NH_2, and NH should be comparable with those observed. However, laboratory studies are required to confirm these speculations.

4.1.2 Very Large Molecules in Diffuse Clouds?

It usually does not take very long for the carrier of a newly discovered line in an interstellar astronomical spectrum to be identified (as in the case of CH, CH^+, and CN, the spectra were discovered in 1937 and, as mentioned above, all of them were assigned by 1941). In most cases, the spectral line is already known from laboratory work, and in some cases the carrier of the line can be predicted from theoretical knowledge of the atom or molecule. In a few exceptional cases, new species need to be constructed in the laboratory so that their spectra can be studied. But usually, in at most a few years, the species that generate interstellar features in astronomical spectra can be identified.

However, there are two series of spectra that are interstellar in origin and have been known for many decades, and yet the carriers of these spectra remain unidentified. This, surely, is a scandal for science and a severe embarrassment to astronomers, chemists, and spectroscopists, in equal measure!

One set is called the Diffuse Interstellar Bands (the DIBs). These are optical absorption lines, rather broader than normal atomic interstellar lines but generally rather weak, and are found on paths through the diffuse interstellar medium towards bright stars that provide the background source of radiation. Several hundred of these features are now known to exist, and the wavelength range in which they occur is from 443 nanometres in the blue end of the optical region (apparently none occur at shorter wavelengths) into the near infrared around one micron. Figure **4.1** shows a representation of the DIB spectrum. This spectrum is unassigned.

The other unidentified interstellar spectroscopic features are called the Unidentified Infrared Bands (the UIBs). These are seen in emission and are almost always associated with bright sources of radiation, such as young hot stars, suggesting that the emissions are excited by radiation. There are just a few of these emission features in the near infrared at wavelengths near 3.3, 6.2, 7.7, 8.7, 11.3, 12.0, and 12.7 microns; see Figure **4.2**.

84   Chapter 4

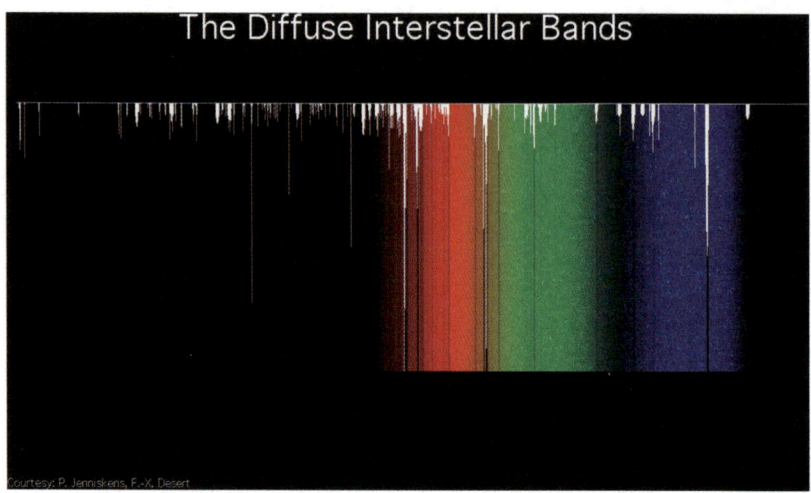

Figure 4.1 A representation of a portion of the spectrum of the DIB, superimposed on the optical spectrum (photo: P. Jeniskens and F.-X. Desert Reproduced with permission).

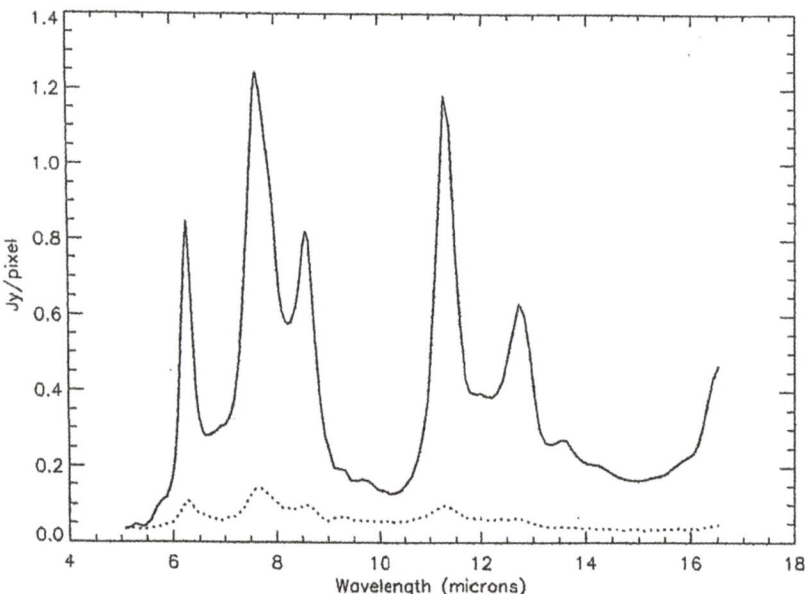

Figure 4.2 Spectrum of the UIBs in the range 5–14 μm at two different positions in the reflection nebula NGC 7023. The features are commonly attributed to vibrational modes of polyclyclic hydrocarbons. Reproduced with permission from D. Cesarsky *et al.* 1996, Astronomy & Astrophysics **315** L305. Copyright ESO.

The features in the UIB spectrum are similar to features arising in organic molecules. For example, the 3.3 micron band is found in organic molecules to arise in the aromatic C−H stretch vibration, *i.e.* where the H is bound to a C-atom in a 'graphitic' structure. The 6.2 and 7.7 micron bands are associated with C−C stretching mode in a 'graphitic' structure, the 8.7 feature with C−H in-plane bending modes, and the 11.3, 12.0, and 12.7 micron features are associated with C−H in-plane out-of-plane bending motions for peripheral mono, duo, and trio H-atoms bound to a 'graphitic-type' structure.

So it was suggested that the carriers of the UIBs were hydrocarbon molecules possessing hexagonal structure, as in graphite. These could either be free-flying molecules, or embedded in amorphous carbon dust grains. The excitation of these emissions required that the carriers be hot. A single molecule—if not too large, say, around 100 atoms—could be made hot enough by the absorption of a single photon of starlight. The molecule would then cool by the UIB emissions and could be re-excited by another photon. These hydrocarbon molecules are of a type known as 'polycyclic aromatic hydrocarbons', or PAHs, so this idea became known as the PAH hypothesis. It should be regarded as a hypothesis because precise fits (in wavelength and width) of the observed UIB features to laboratory data for specific PAHs remain to be established.

Nevertheless, it seems reasonable to assume that PAHs do exist in the diffuse interstellar medium and that the problem of identifying specific carriers is that PAHs of a particular size can have a huge variety of structures. Perhaps interstellar PAHs exist but the specific PAHs used in laboratory experiments are not among them. If one asks how such large molecules could be formed in interstellar space, then one finds that it is difficult for such large structures to be built up from smaller ones, as the smaller structures (say, less than about 50 atoms) tend to be destroyed by the radiation field or by the damaging effects of helium ions or oxygen atoms.

An alternative origin may be that PAHs arise as a result of the shattering of carbon grains in collisions with each other or with other species. If so, it is not surprising that interstellar PAHs do not match perfectly the spectra of specific hydrocarbons in the laboratory, such as naphthalene (the simplest PAH, a pair of fused benzene rings) or anthracene (three fused benzene rings in a straight line)—both commonly used in making comparisons between laboratory and astronomical data. In fact, observations suggest that these two species probably do not exist in the diffuse interstellar gas.

The assignment of the DIB features is proving to be much more difficult than that of the UIBs, for which the basic attribution to carbonaceous material is probably correct. With DIBs, we do not have any similar clue as to the nature of the material responsible for these absorptions. Even though the DIBs are much more numerous than the UIBs, and we can see them in the spectra of interstellar matter along the lines of sight to many bright stars, so there is no shortage of spectroscopic data, it is hard to find any observational or theoretical pattern that might suggest an origin.

However, the idea that PAHs of a significant size may well be present in interstellar space does suggest that PAHs may also be associated with the DIBs. A DIB absorption line can then be interpreted as an electronic transition in the PAH molecule, and the breadth of the line can be attributed to rotational structure of the molecule. Theoretical studies of such absorptions from large molecules suggest that the width of the absorption line is related to the size of the molecule, and that—at least for a couple of the DIB absorptions—molecules of about one hundred atoms are required, similar to the ideas we have from considering the UIBs. So, it is possible that DIBs and UIBs may be related phenomena. However, the precise molecules involved—assuming that they arise from the shattering of grains—are quite unlikely to be identical to those we find in bottles on laboratory shelves.

The PAHs, if they exist in the interstellar medium, do have an effect on interstellar chemistry. They tend to 'mop up' electrons very easily, and will become negatively charged in interstellar clouds. These charged PAHs move much more slowly than free electrons in the gas phase. This means that positive ions in the gas, such as HCO^+, or H_3O^+, do not encounter free electrons as often as they would otherwise, so the rate of loss of positive ions declines and the abundances of positive ions become larger. These effects are more important in denser, darker clouds, which we discuss in the next Section. However, we do not have any information at all about DIBs in denser, darker clouds, because the starlight is too heavily extinguished for us to be able to see the weak DIBs.

4.2 DARK CLOUDS

Dark clouds are colder, denser, and—most importantly—darker than diffuse clouds (see Table **1.2**). They may be isolated clouds such as Barnard 68 (Figure **1.2**), containing just a few hundred solar masses of material, or vast accumulations of gas, up to a million solar masses, in the form of giant molecular clouds such as those in our own galaxy, the Milky Way, and in other galaxies such as the one shown in Figure **1.4**. In general, as the densities are higher than in diffuse clouds, they also contain more dust, and so they are darker. They have optical depths in the visible of more than about five, meaning that the intensity of optical starlight inside the cloud is less than one percent of its mean intensity outside the cloud, and in the ultraviolet, where dust is much more strongly absorbing, the intensity may be reduced by a factor on the order of 10 000. Since it is the ultraviolet component of starlight that is usually important in driving chemistry in interstellar gas, we conclude that starlight is not important for chemistry in the interiors of dark clouds, and that cosmic rays are important drivers of the chemistry. That part of the chemistry in diffuse clouds that is driven by cosmic rays also applies directly in dark clouds.

In particular, the lack of ultraviolet starlight in dark clouds means that hydrogen molecules are only destroyed by cosmic rays or as a consequence of reaction sequences initiated by cosmic rays. This destruction is a much slower process than destruction by unshielded starlight, but since the formation of H_2 on dust grains continues as efficiently as in diffuse clouds the balance in hydrogen is strongly molecular, rather than atomic. In fact, as discussed in Section **2.4**, in steady state (which may not always be achieved in the Milky Way galaxy) the number density of H-atoms in dark clouds is about 1 per cm^3, while the number density of H_2 molecules may be several thousand times larger. The consequence for chemistry in dark clouds is that routes driven by H_3^+ are much more important than those driven by H^+.

4.2.1 Gas Phase Molecules in Dark Clouds

Observations show that CO molecules are by far the most abundant molecules in dark clouds of the Milky Way galaxy, after H_2 molecules. In fact, much of the available carbon that is not incorporated into dust grains is in CO molecules. Since there are about 3 C-atoms for every 10 000 H-atoms (see Section **1.2.1**), and some of those are locked up in carbon dust grains, there is—very roughly—about one CO molecule for every 10 000 H_2 molecules in dark clouds.

This suggests that we could use CO emission as a tracer of H_2. Molecular hydrogen is abundant but curiously difficult to observe, since it has no transitions that can be excited in low temperature (~ 10 K) gas. But almost all the mass of gas in a cloud is in molecular hydrogen, so perhaps we could use CO emission as a proxy for cloud mass. That would be very useful, since emission by CO molecules in the rotational transition 1−0 is easy to detect. However, emission by CO molecules in the interior of a cloud can be absorbed by CO molecules nearer the edge of a cloud, so one expects that the CO emission will not rise in proportion with the cloud mass, but is less intense than naively expected for larger clouds.

In fact, observations seem to show that, on average, we do find that cloud mass is proportional to CO emission intensity. One interpretation of this surprising result is that the clouds are clumpy, and that the CO emission 'counts the clumps', *i.e.* rises in proportion to the number of clumps in the cloud. If so, then we can still use the CO emission as a measure of the mass, but we shall also want to look at the chemical consequences of having a range of masses in a cloud.

Internal structure in a cloud may be important. Denser parts of a cloud may be those regions that are undergoing collapse due to gravity, and may be on their way to forming stars. But it is difficult to observe structure within a cloud, because single-dish telescopes do not have enough angular resolution. But if one links a number of dishes together then the effect is to create a telescope of size comparable with the array of telescopes; this has much greater resolving power. Figure **4.3** shows contour diagrams obtained in this way of a portion of an isolated dark cloud called L673 for emission from two molecular species; the formyl radical ion (HCO^+) and carbon monosulfide (CS). These observations show

Molecules in the Milky Way Galaxy 89

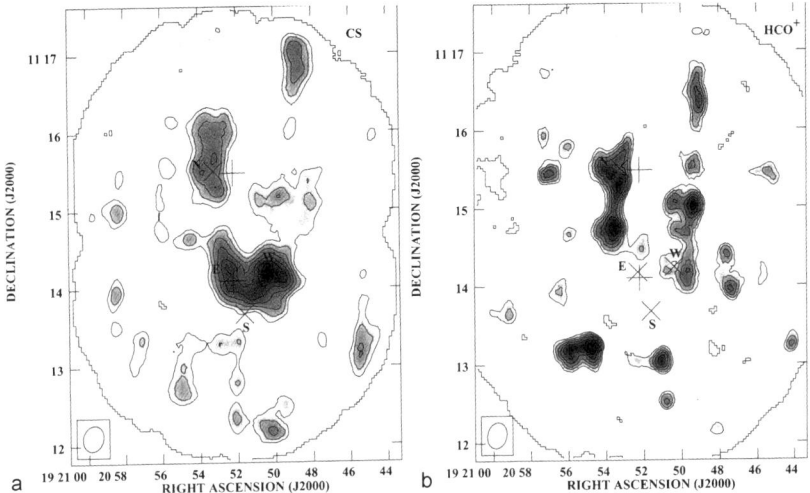

Figure 4.3 Contour maps of the intensity of emission from CS and HCO$^+$ molecules in a portion of the dark cloud L673. Evidently, the dark cloud is clumpy, but the clumps are not the same in the emission from both molecular species (a point to which we return in Chapter **8**). Reproduced with permission from O. Morata *et al.* 2003, Astronomy & Astrophysics **397** 181. Copyright ESO.

the existence of clumps very much denser than the ambient gas in which they are embedded, and the largest of them is quite massive, containing about a solar mass. It is possible that its internal pressure may be unable to support its weight, and that it may collapse and possibly form a star. We will discuss star formation in the next section.

These tracer molecules are formed by the processes similar to those discussed for diffuse clouds. Carbon monosulfide (CS) forms easily in neutral exchange reactions such as that between CH and S. The formyl radical ion is formed similarly in rapid reactions of CO with H_3^+, and N_2H^+ is formed similarly.

The molecules are very useful tracers of high-density gas. Other frequently used tracers of dense gas in dark clouds are hydrogen cyanide (HCN), hydrogen isocyanide (HNC), and ammonia (NH$_3$), and N_2H^+ is formed similarly. The cyanides and isocyanides are created by a number of ion-molecule and neutral exchange reactions (see Section **3.3**).

The tracer molecule ammonia may be made on surfaces (see Section **4.1.3**) or through a long process involving nitrogen ions released from nitrogen molecules (or other nitrogen containing molecules) by helium ions in dissociative charge exchange:

$$N_2 + He^+ \rightarrow N^+ + N + He$$

where the He^+ ions are the result of cosmic rays striking helium atoms. The N^+ ions react with hydrogen molecules, abstracting a hydrogen atom in each interaction:

$$N^+ \rightarrow NH^+ \rightarrow NH_2^+ \rightarrow NH_3^+ \rightarrow NH_4^+$$

The nitrogen ion can add no more than four hydrogen atoms, which then dissociatively recombines with an electron:

$$NH_4^+ + e \rightarrow NH_3 + H$$

to release ammonia.

4.2.2 Molecules in the Solid Phase—Increasing the Molecular Complexity

Observations in the infrared along lines of sight through sufficiently dark clouds reveal the existence of ices deposited on dust grains, and we have seen that the chemical composition of these ices points to the operation of an active surface chemistry (see Sections **3.4.2** and **3.4.3**). It is not clear what triggers the deposition of the ices; the critical depths for deposition are not the same for all dark clouds. However, the most important issue for astrochemistry is that—beyond this critical depth into a cloud—molecules and atoms are being lost from the gas in collisions with dust grains. If this process is unrestricted, then these ices will take up almost all the gas phase species apart from hydrogen molecules and helium atoms. This provides a natural timescale for chemistry and other events in molecular clouds—unless a mechanism operates that actively returns molecules from the ice to the gas. If not, the freezing-out of atoms and molecules from the gas occurs (for Milky Way conditions) in a timescale that is about one million

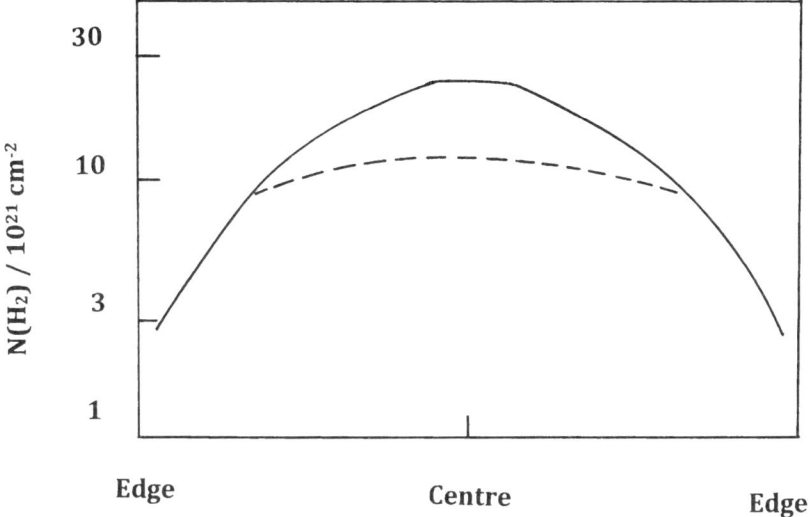

Figure 4.4 The column density of molecular hydrogen along a diameter of a near-spherical pre-stellar core, L1689B, is obtained from observations in two independent ways. Emission at 850 microns is attributed to dust grains in the core, and assuming a canonical dust: gas ratio allows an estimate of the abundance of gas (as H_2) to be made; this is the solid line. The second method measures the CO ($J = 2 - 1$) emission intensity. With a standard CO : H_2 ratio, a second estimate of the H_2 abundance can be made; this is the dashed line. The discrepancy is a measure of the 'missing' gaseous CO that has frozen out on to the grains. Redrawn from M. Redman *et al.* 2002, Monthly Notices of the Royal Astronomical Society **337** L17. Copyright Royal Astronomical Society.

years for many dark clouds. Thus, gas phase chemistry really has to occur within this timescale—unless there is a rapid return of material from ice to gas. So we can expect clouds that are 'old' to have lots of ice and rather few molecules in the gas phase; see Figure 4.4.

These ices are important in interstellar chemistry. Not only does their growth limit the time available for gas phase chemistry in interstellar clouds, the ices also provide an opportunity for solid-state chemistry to occur. This may need to be initiated by cosmic rays, by starlight, or by warming. Such chemistry tends to lead to greater chemical complexity, since the reactants in the solid are at very high density, rather than the extremely low densities of interstellar clouds.

So the detection of methanol, formaldehyde, and carbon dioxide leaves no doubt that the chemical processing of ices is occurring in interstellar clouds. However, a further and more dramatic chemical enrichment of the gas phase occurs during the evaporation of these ices near to a newly formed star, when the stellar infrared radiation warms the previously cold ice-laden dust grains deep inside a dense star-forming core. This stellar infrared radiation 'reaches the parts that other stars cannot reach', to coin a phrase. Since these warming and evaporation processes are an essential part of star formation, we will discuss them and other issues in the following Section.

4.3 STAR FORMATION

Star formation is one of the most dramatic events in astronomy. The image on the cover of this book is of a huge star-forming region in the Large Magellanic Cloud, a neighbouring galaxy to the Milky Way. Star-forming regions show great activity. Chemistry plays an important part in initiating that activity and tracing its progress.

In the Milky Way galaxy, new stars appear at a rate of about one a year. They form from the interstellar gas in dark clouds. This conversion of gas to stars requires that dark cloud gas, in objects known as dense cores, with densities of about ten thousand to one hundred thousand hydrogen molecules per cm^3 be compressed by gravity to the density of a star, which—for our Sun—is about 1 g per cm^3, or roughly 10^{24} H-atoms per cm^3. Therefore, the conversion requires a density increase by roughly 20 orders of magnitude implying a collapse in radius of a spherical cloud by a factor of around ten million.

The way in which this conversion process occurs is one of the main themes of modern astronomy. Fortunately, molecules are present throughout the early stages of this transformation from interstellar gas to 'proto-star' (*i.e.* before stellar thermonuclear processes begin) and these molecular emissions can help astronomers to understand how star formation occurs. The emissions, particularly in the early stages of the collapse, are also important in cooling the gas.

For the temperature is a key parameter in this process. Compression of a gas will tend to heat the gas, and a higher temperature means a higher pressure, tending to restore the gas to its previous state. But if heat can be radiated away from the gas, then the temperature may not rise, and the pressure may remain insufficient to resist the compression so that the collapse may continue. So molecules can play several key roles in star formation;

firstly, as tracers—through their emissions we can trace the physical conditions in the gas, and secondly, as coolants—their emissions allow gravity to continue to compress the gas. A third key role involves the effect of the chemistry controlling the fractional ionisation and—through it—the dynamical influence of the magnetic field. The magnetic field can resist the collapse driven by the gravity and transport angular momentum, if the fractional ionisation is high enough that collisions between ions and neutrals occur frequently.

As the gravitationally driven compression continues, there comes a point at which the force of gravity becomes so strong that a big rise in temperature—and therefore in pressure—can occur without disrupting the proto-star. Molecules near the star are then dissociated and atoms ionised. So the role of molecules in star formation ceases at that point.

4.3.1 Chemical Complexity in Star-Forming Regions

Towards regions of formation of stars of both low and high mass, newly formed stars are found in association with very dense tiny clumps of relatively warm gas. These so-called 'hot cores' and 'warm cores' are presumed to be material that was part of the collapsing cloud, but which wasn't incorporated into the proto-star. Figure **4.5** shows the relation of a hot core to the young star with which it is associated. By interstellar standards, the density of these hot and warm cores is exceptionally high, typically tens of millions of hydrogen molecules per cm^3, *i.e.* thousands of times denser than the dark clouds from which they originated. They are warmer, too, typically a few hundred K (near high mass proto-stars, but cooler near proto-stars of more modest mass), rather than the typical 10 K of dark clouds. The rise in temperature is caused by the close proximity of the proto-star that is emitting powerfully in the infrared.

The exceptional feature of these dense cores in star-forming regions is their chemical complexity in large organic molecules, observed in the gas phase. As we have mentioned in Chapter **1**, they contain a wide variety of species, (see Table **4.2** for a list of large organics detected near the centre of the Milky Way galaxy).

Gas phase routes to form these species simply are not efficient enough to produce these relatively complex molecules, and it is accepted—and supported by many laboratory experiments—that only chemical processing of the relatively simple ices can generate

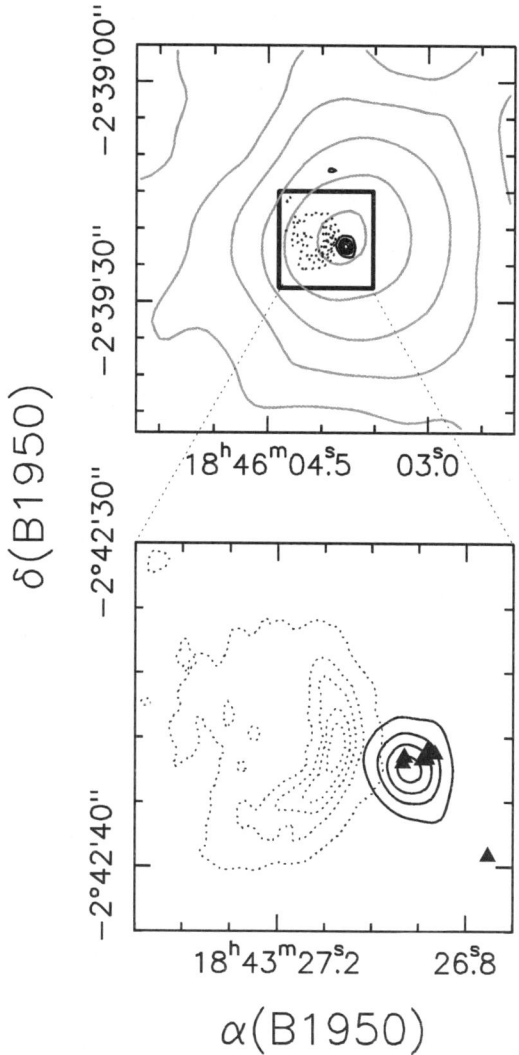

Figure 4.5 The distribution of CS and CH$_3$CN molecules in a hot core. The upper panel shows that CS (grey contours) is widely distributed around the hot core region, while the lower panel shows that CH$_3$CN (black contours) is concentrated close to the ionised region containing the newly forming star. The ionised region is traced in radio emission at a wavelength of 1.3 cm (dotted contours). The triangles show the position of H$_2$O masers, which are regarded as signatures of star formation. (Courtesy: R. Cesaroni; unpublished figure).

Table 4.2 Some of the larger molecular species that have been detected in the centre of the Milky Way galaxy. Undoubtedly, many more species are present and detectable.

Five atoms: HCOOH (formic acid), HC_3N (cyanoacetylene), H_2CCO (ketene)
Six Atoms: CH_3OH (methanol), CH_3CN (methyl cyanide), CH_3NC (methyl isocyanide), HC_2CHO (propynal)
Seven Atoms: CH_3CHO (acetaldehyde), CH_2CHCN (vinyl cyanide), c-C_2H_4O (ethylene oxide)
Eight atoms: $HCOOCH_3$ (methyl formate), CH_2OHCHO (glycolaldehyde), CH_3COOH (acetic acid), CH_2CHCHO (propenal)
Nine Atoms: CH_3OCH_3 (dimethyl ether), CH_3CH_2CN (ethyl cyanide)
Ten Atoms: CH_3COCH_3 (acetone), $(CH_2OH)_2$ (ethylene glycol)

---

this complexity. It is argued that these complex organics are formed in or on the ice mantles on dust, and then desorbed into the gas phase when the grains are warmed by the nearby proto-star. How does this complexity arise?

4.3.2 Routes to Chemical Complexity in Star-Forming Regions

A plausible scenario in which simple ices might generate more complex species has been developed by Robin Garrod (at Bonn) and colleagues Susanna Widicus Weaver (at Illinois) and Eric Herbst (at Virginia). It starts from the recognition that there is a weak radiation field present even inside the dense cores within dark clouds. This weak field is initiated by the cosmic ray ionisation of molecular hydrogen; then collisions between the electrons released in these ionisations and hydrogen molecules excite the molecules, and finally the subsequent decay from these excited levels creates a weak ultraviolet radiation field. This field is capable of dissociating simple molecules of the original ice, *i.e.* water (H_2O), methane (CH_4), formaldehyde (H_2CO), methanol (CH_3OH), and ammonia (NH_3), creating radicals such as H, OH, CO, HCO, CH_3O, CH_2OH, CH_3, NH, and NH_2. In general, these radicals are not mobile on the ice at a temperature of about 10 K, so the radicals remain bound. But the heat of the nearby proto-star warms these dense cores; the warming occurs rather slowly, perhaps over thousands of years, to the temperatures of several hundred K that are observed. During the warming process, some radicals become mobile before others, and react to create new species; then as the warming process continues more massive and more strongly bound radicals may eventually become mobile. So, in this picture, the actual chemistry occurring depends on the particular stage in the warming process.

So, for example, early in the warming process acetaldehyde (CH_3CHO) may form from the combination of weakly bound radicals CH_3 and HCO, while later in the process ethylene glycol [$(CH_2OH)_2$], might form from the combination of two more strongly bound CH_2OH radicals. Evidently, the warming and the final temperature may be different in the case of high-mass and low-mass proto-stars, and this difference may give rise to chemical differences in cores near these types of object.

Is there any laboratory evidence in support of this kind of picture? Yes, there is a very large and continuing programme of relevant experiments in various laboratories. One important series of experiments conducted at Leiden by Karin Öberg and colleagues involves the deposition of specific ices (either pure or of mixed composition) in ultrahigh vacuum and at low temperature. The ices are irradiated by ultraviolet radiation from a hydrogen discharge lamp for a measured dose, and the ices slowly warmed. A Quadrupole Mass Spectrometer detects gas phase molecules released during the warming process and infrared spectroscopy is also used as a tool in the identification process. These experiments have been so successful that they are now moving beyond the qualitative phase into quantitative work; *i.e.* determining the ratios of one product molecule to another, where both are formed in this solid-state chemistry (See Section 9.4).

Experiments at Leiden involving the irradiation of methanol ice and methanol/carbon monoxide ice produced all the C-, O-, and H-bearing complex organics commonly observed in star-forming regions. Some of these species are also formed in $H_2O : CO_2 : CH_4$ ices. Nitrogen-containing species require NH_3 or CN-containing species within the ice. The proportion of water in the ice is found in these experiments to affect the diffusion barriers of hydrogen-bonding radicals relative to non-hydrogen-bonding radicals. The chemical evolution of the ice therefore is affected by the chemical nature of the initial simple ice.

Therefore, the evidence strongly supports this kind of picture of ice processing. It is certainly well established that chemical complexity can be induced in a simple ice by irradiation. The experiments are very convincing. But—from a theoretical point of view—many of the details remain uncertain at this stage. The mobility of radicals across the grains, the binding energies of species to mixed ice grain surfaces are poorly known, and the desorption process from a mixed ice is more complex than normally envisaged.

Desorption from mixed ices does not seem to be the simple process that one might imagine. A study by Martin McCoustra and colleagues (then at Nottingham) of desorption from an irradiated mixed ice of carbon monoxide and water when slowly warmed (*i.e.* a temperature programmed desorption experiment, or TPD) from very low temperature reveals that desorption of CO molecules from the ice occurs in four discrete and narrow temperature bands, rather than a single broad range of temperature above some critical value. First, CO desorbs from pure solid CO, at a temperature of about 25 K. Then, at a slightly higher temperature, CO desorbs from a monolayer bound to surface water molecules. Next, a restructuring of the water ice from amorphous to crystalline releases CO trapped within the ice (the so-called 'volcano' effect); and finally, water ice itself desorbs along with any remaining CO (so-called 'co-desorption'). Carbon monoxide molecules are ejected at each of these critical temperatures, not just at the single temperature associated with pure solid CO. Other molecules behave in a similar way, though not necessarily with four distinct temperatures of evaporation (see Figure **4.6**). Therefore, each molecule that is desorbed will appear at one or more particular moments during the warming process.

Another complication not yet resolved is the feasibility of storing radicals in a cold ice. As shown by the astrochemist Mayo Greenberg, it is possible to store a population of radicals in a cold ice, for long periods, up to a critical density of radicals that depends on the temperature. In experiments in his Leiden laboratory on ices with a population of stored radicals, he found that raising the temperature from very low values caused explosions to occur at a critical temperature, evidently triggered by a runaway process. For a specific density of radicals in the ice, the first recombination releases energy that triggers a runaway process of recombination among all the stored radicals, and an explosion occurs.

Such explosions have also been observed in many other experiments (see Section **2.2.3**), some of which may raise the temperatures of the dust grains to values on the order of a

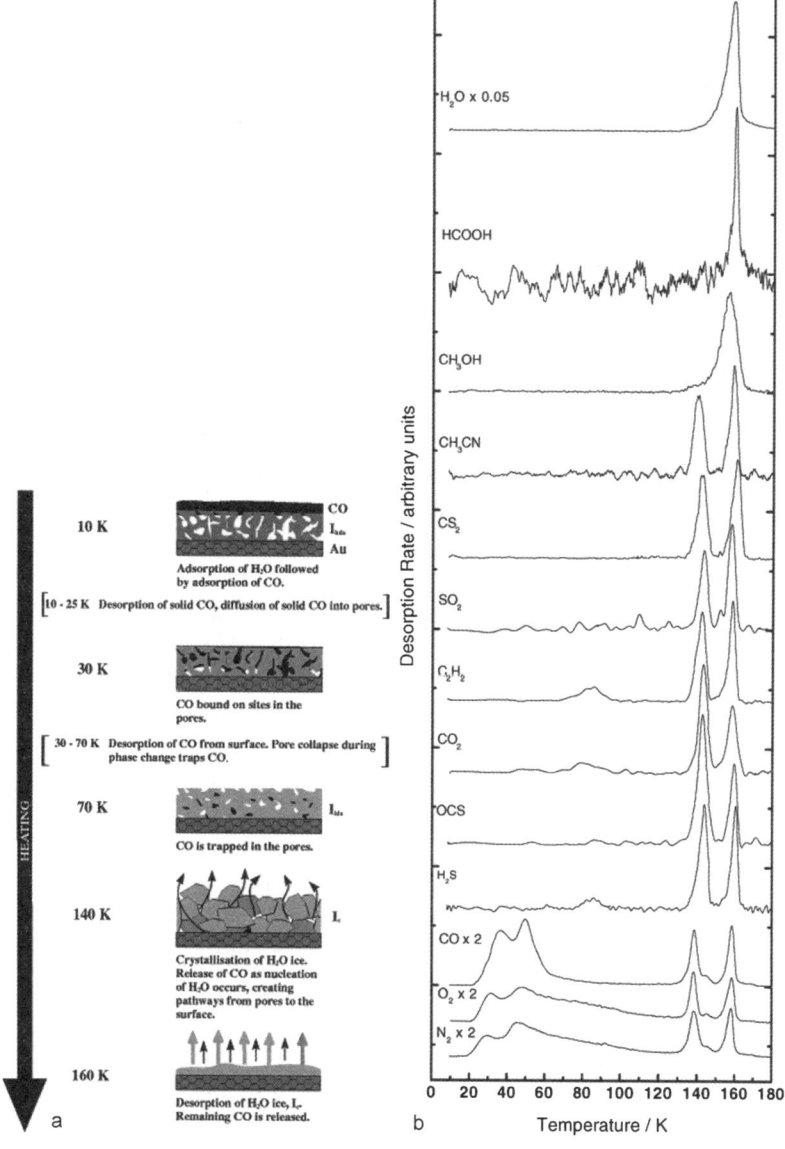

Figure 4.6 (a) Cartoon illustrating schematically the TPD behaviour of CO/H_2O ice. Reproduced with permission from M. P. Collins *et al.* 2003, Astrophysical Journal **583** 1058, Copyright American Astronomical Society. (b) TPD results for a variety of molecular species on H_2O ice. Reproduced with permission from M. P. Collins *et al.* 2004, Monthly Notices of the Royal Astronomical Society **354** 1133. Copyright Royal Astronomical Society.

thousand K. So it is quite possible that explosions in radical-containing ice mantles on dust grains occur before the warming processes allow greater mobility. If so, then the chemistry will occur in the explosions, rather than in the gentle warming process. Such ideas are currently under investigation.

4.3.3 Deuterium Fractionation

The cosmic abundance of deuterium relative to the main isotope, hydrogen, is low, about 1.6×10^{-5}, so it was a surprise when observations of some molecules and the corresponding deuterated isotopologue were made in hot cores, and the abundance ratio of the two species was very much larger than the cosmic D : H ratio. For example, in the hot core in the star-forming region of the Orion constellation the measured ratio of $HDCO : H_2CO$ is about 0.14, of $CH_3OD : CH_3OH$ is about 0.03, and of $C_2D : C_2H$ is about 0.05. Evidently, it is a general phenomenon, not limited to one or two species. This concentration of deuterium at the expense of hydrogen in molecules, sometimes by several orders of magnitude, is called fractionation. It is found to occur not only in warm hot cores, but also in cold dark clouds as well. Fractionation is observed not only for deuterium, but also for isotopes of carbon and of oxygen.

Where is the deuterium in dark clouds? Since the formation processes for HD and for H_2 are similar for the H and D isotopes, nearly all of the deuterium is in HD. Therefore, this is the reservoir of deuterium in interstellar clouds.

What is the mechanism by which it is concentrated in molecules? There is a slight difference in the energy levels of HD and H_2. The greater mass of deuterium means that the ground state is very slightly lower for HD than for H_2, so in reactions where both HD and H_2 are involved, HD is favoured at low temperatures. For example, in the reaction:

$$H_3^+ + HD \leftrightarrow H_2D^+ + H_2$$

the products on the right hand side are slightly lower in energy, by an amount equivalent to 178 K. Above this temperature, the reaction can go in both forward and back directions, but in gas with temperature below 178 K the back reaction is suppressed while the forward reaction channel is open. Therefore in cold clouds, with temperatures well below this critical temperature, this forward reaction tends to favour H_2D^+ over H_3^+ and high fractionations will occur, given enough time.

Given the importance of H_3^+ in cold cloud chemistry, it is easy to see that an overabundance of H_2D^+ can lead to an over-population of deuterium in other species. For example, the reactions:

$$H_2D^+ + H_2CO \rightarrow H_2DCO^+ + H_2$$

$$H_2DCO^+ + e \rightarrow HDCO + H$$

tend to populate deuterium in the formaldehyde molecule. Of course, there are other possibilities, too. For example, in the dissociative recombination reaction above, H could be retained and D ejected. But if that were the case, the formaldehyde would be unchanged. It is only the case in which D is retained that the fractionation is enhanced.

There are other fractionating mechanisms, too, but they all depend on the small energy difference between the isotopologue and the main molecule.

Why is deuterium fractionated in hot cores, where the temperature is relatively high? Typical temperatures may be in the range 200–300 K. This is above the critical temperature for fractionation, discussed above. The reason is that the hot core matter has spent a long time at low temperatures, around 10 K, before it was heated by the proto-star. It was in that long evolutionary period that high fractionations could be established, and the higher temperatures make the molecules more easily detectable.

4.4 NEAR-STELLAR ENVIRONMENTS

The surface temperatures of most stars are too hot and their radiation fields too harsh for molecules to exist in abundance in their atmospheres. The Sun's surface temperature is about 5800 K, and molecules do not exist in its atmosphere—or not for very long. However, (as pointed out in Chapter 1) in sunspots the temperature is rather lower at about 4000 K. Detections have been made of ultraviolet emissions from molecular hydrogen (H_2) and carbon monoxide (CO) in sunspots. Evidently, the Sun is a critical case, and, in general, stars with surface temperatures lower than that of the Sun are expected to contain molecules. Cool stars with surface temperatures as low as, say, one or two thousand K are expected to be entirely molecular in their surface layers.

The number densities in the atmosphere of a cool star are very high (around 10^{12} H_2 molecules per cm^3) and given that the temperatures are more than a thousand K, and that there is no ultraviolet radiation field that might photodissociate molecules, the chemical composition is simply determined by rearrangement collisions, in which the molecules arrange themselves into the most stable species for the given elemental abundances, subject to the prevailing density and temperature. This lowest energy arrangement is said to be in local thermodynamic equilibrium, or LTE.

For cool stars that have more carbon in their atmospheres than oxygen, then—as mentioned in Section 3.6—the LTE chemistry preferentially ties up almost all the oxygen in CO, and the excess carbon appears in acetylene (C_2H_2) and the ethynyl radical (C_2H), methane (CH_4), hydrogen cyanide (HCN), along with nitrogen (N_2) and ammonia (NH_3), and a few other species. For cool stars that have more oxygen than carbon, then LTE chemistry ties up almost all of the carbon in CO_2, and the rest of the oxygen appears in water (H_2O), carbon dioxide (CO_2), nitrogen monoxide (NO), silicon monoxide (SiO), and various other oxides. These molecules will be the 'parent' molecules from which a new chemistry will form 'daughter' species, later in the evolution of the wind chemistry.

4.4.1 Dust Formation

Some cool stars are seen to dim and brighten again, repeatedly. In some cases, this is interpreted as due to dust formation in the outflowing envelope just above the atmosphere. As the envelope expands, the density falls and the dust extinction is therefore diluted. Then dust formation re-occurs.

Our understanding of the formation of carbonaceous dust is based on ideas from combustion chemistry. The problem is one of conversion of the simple species acetylene and ethynyl radical available in the stellar atmosphere into an amorphous solid similar to soot. The process is considered to take place in two stages; firstly, nucleation and, secondly, condensation. In the case of carbonaceous dust formation, nucleation is the formation of large organic molecules, *i.e.* PAHs, in the gas phase. Condensation is the association of these large molecules into three-dimensional solids.

One possible scheme based on combustion chemistry is illustrated in Figure **4.7**. This suggests that the crucial step—making the first ring structures—involves the reaction of two three-member hydrocarbons (propargyl, C_3H_3) or a 4-member hydrocarbon (such as 1-buten-3-ynyl, C_4H_3) with acetylene. The growth from a ring to full PAH status may be achieved through acetylene addition to the ring, hydrogen abstraction to create an aromatic radical, and the addition of a second acetylene and ring closure.

All these reactions are reversible, and so there are constraints on the physical conditions under which the desired nucleation can occur. The models predict a rather narrow temperature range of a few hundred K around a temperature of about 1000 K for nucleation. Clearly, not all stars will meet these conditions.

Condensation is envisaged as beginning with PAH-dimer formation, where the aromatics are bound by weak van der

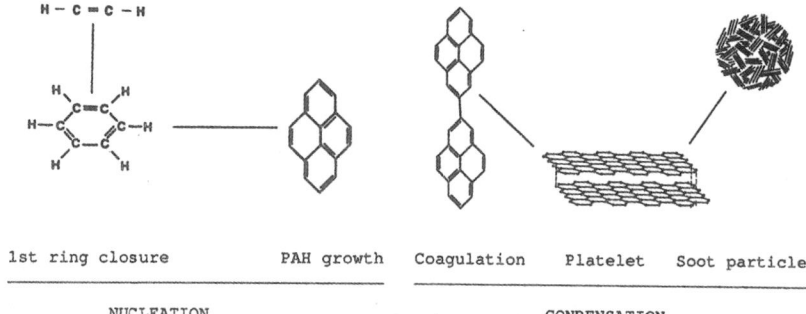

Figure 4.7 Possible mechanisms for the formation of organic molecules in cool circumstellar envelopes. Reproduced with permission from I. Cherchneff in Molecular Astrophysics of Stars and Galaxies, eds. T. W. Hartquist and D. A. Williams 1998, Clarendon Press, Oxford, p.265.

Waals forces. The condensation would then continue through the coagulation of condensation products, and the accumulation of acetylene on the carbonaceous solid. A possible scheme is illustrated in Figure **4.8**.

The formation of dust grains in oxygen-rich environments is less well characterised. The crucial step is, as in the carbon-rich case, the nucleation. Recent work suggests that nucleation from SiO monomers does not occur, but that heteromolecular nucleation involving Mg, SiO, and H_2O may proceed.

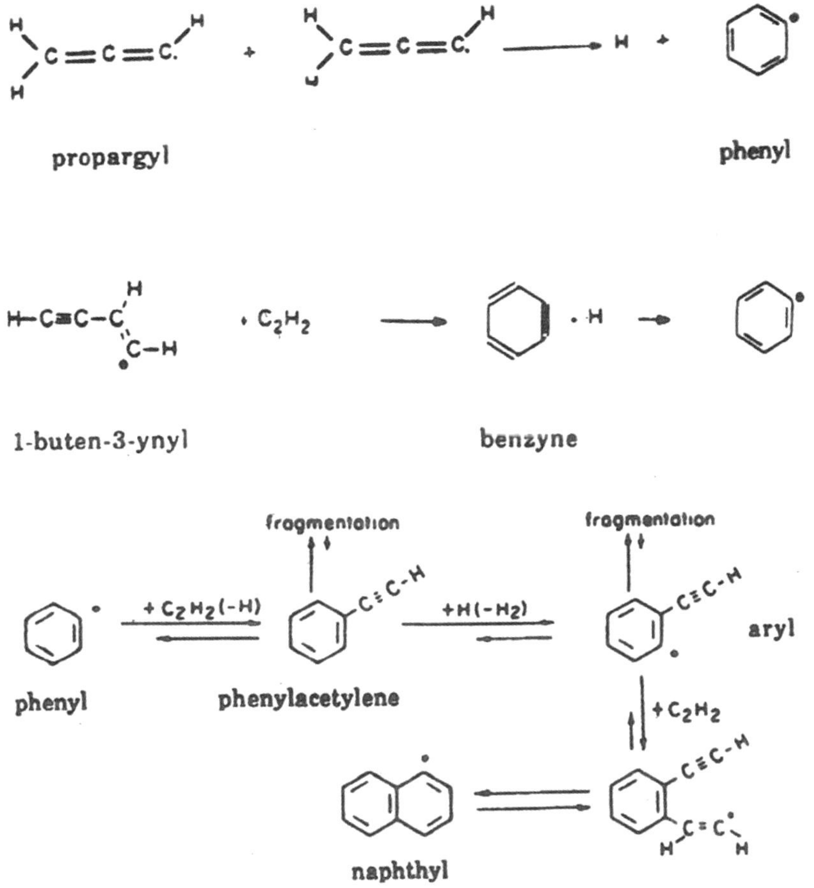

Figure 4.8 Possible mechanisms for the formation of solid carbonaceous material in cool stellar atmospheres. Reproduced with permission from I. Cherchneff in Molecular Astrophysics of stars and galaxies, eds. T. W. Hartquist and D. A. Williams 1998, Clarendon Press, Oxford, p.265.

4.4.2 Chemistry in Cool Circumstellar Envelopes

Stars of both types (carbon-rich and oxygen-rich) are observed to develop winds, which drift out into interstellar space with velocities of 10–20 km s^{-1}. Radiation pressure on dust formed just above the stellar atmosphere may help to drive the wind. But as the gas flows out from the star, its density becomes lower and the extinction caused by the dust that is embedded in the gas becomes weaker, so that eventually ambient starlight from interstellar space begins to penetrate the envelope and to ionise and dissociate 'parent' molecules, *i.e.* the molecules from the stellar atmosphere that have not been incorporated into dust. A characteristic photochemistry is then initiated, forming new 'daughter' species from the original 'parent' species. Eventually, however, the extinction becomes so weak that the interstellar radiation field destroys the 'daughter' species, so that they exist only in a fairly well-defined shell around the star.

For example, observations of CS and SiS molecular emissions near a well-studied cool star show that these species are confined to a region close to the star, while emissions from CN and SiC$_2$ appear to arise in shells around the star, outside the CS and SiS emissions. Thus, CS and SiS are considered to be "parent" species and CN and SiC$_2$ are "daughter" species.

The chemistry that produces these 'daughter' molecules is driven by starlight, and is based to some extent on reactions between ions and molecules that produce new ions, followed by recombination of these new ions with electrons. Neutral exchange reactions can also be important. However, although the envelope chemistry is driven by starlight—as is chemistry in diffuse clouds—it is different from diffuse cloud chemistry in various ways. In stellar envelopes, the chemistry begins with the 'parent' species, while in diffuse clouds the chemistry begins with molecular hydrogen in reaction with atoms and atomic ions. Of course, the density in stellar envelopes is much higher than arises in diffuse clouds. Also, stellar envelopes represent a dynamical situation in which molecules formed in a parcel of gas moving with the stellar wind survive only for a short time. Diffuse clouds are normally modelled as being static and with the chemistry in steady state.

The richest chemistry occurs in the envelopes of carbon-rich stars (see Table **4.3**). The big surprise in the observations is that

Molecules in the Milky Way Galaxy 105

Table 4.3 Some detections of molecular species, including anions, in the cool carbon-rich envelope of the star IRC +10216. In each category, the molecules are listed in very rough order of abundance (most abundant first). Where necessary, linear (l) and circular (c) forms of a molecule are indicated.

Oxygen-containing molecular species	CO, H_2O, OH, H_2CO, C_3O, HCO^+
Carbon and hydrocarbon species	C_2H_2, CH_4, C_2H, C_4H, C_2, C_3, C_5, l-C_3H, C_6H, C_5H, c-C_3H_2, CH_3C_2H, c-C_3H, C_2H_4, H_2C_4, C_8H, C_7H, H_2C_6, C_6H^-, H_2C_3, C_8H^-, C_4H^-
Nitrogen-containing species	HCN, NH_3, CN, HC_3N, C_3N, HC_5N, HNC, CH_3CN, HC_7N, HC_9N, CH_2CN, HC_2N, C_5N, HC_2NC, C_2H_3CN, C_5N^-, HC_4N, C_3N^-, HNC_3
Silicon-, sulfur-, and phosphorus-containing species	SiS, SiC_2, CS, SiH_4, SiO, C_2S, SiC, C_3S, HCP, H_2CS, SiN, PH_3, H_2S, c-SiC_3, SiC_4, SiCN, C_5S, PN, C_2P
Metal-containing species	AlCl, MgNC, AlF, NaCN, NaCl, AlNC, MgCN, KCl

long carbon chain molecules are prominent in envelopes, more so than in the interstellar medium. The reason is that the gas densities—though falling—are still much higher than in most interstellar clouds, so formation reactions proceed much more quickly in envelopes than in interstellar clouds. We will describe how the carbon chains are formed.

Starlight ionises and dissociates the 'parent' molecule acetylene to form $C_2H_2^+$ and C_2H. Then ion-molecule reactions such as:

$$C_2H_2 + C_2H_2^+ \rightarrow C_4H_2^+ + H_2 \text{ or } C_4H_3^+ + H$$

followed by recombination of ions with electrons to give neutral products which are carbon chains:

$$C_4H_2^+ + e \rightarrow C_4H + H$$

$$C_4H_3^+ + e \rightarrow C_4H_2 + H$$

Ethynyl (C_2H) can undergo neutral reactions:

$$C_2H + C_2H_2 \rightarrow C_4H_2 + H$$

and similar addition reactions can produce larger polyacetylenes. Nitrogen can be inserted into the polyacetylenes by addition with the parent CN radicals:

$$CN + C_{2n}H_2 \rightarrow HC_{2n+1}N + H$$

by direct insertion of nitrogen atoms into carbon chains, or by addition of HCN molecules to those chains.

Models based on these kinds of chemical networks have been successful in accounting for the range of species detected and the distance from the central star at which they show peak abundance. These models also give very precise information on density and radiation field in the wind, and on elemental abundances in the atmosphere.

4.4.3 Chemistry in Planetary Nebulae

Obviously, a cool star cannot supply material to an expanding envelope forever. Eventually, the supply ceases and the envelope becomes detached from the star. At the same time, these changes affect the evolution of the star which by this stage has used up all its fuel (hydrogen): without a thermonuclear power source at its centre it contracts under gravity and becomes much hotter, evolving towards the status of a white dwarf star. As it does, it also sets in motion a much hotter and much faster wind that impinges on the (now detached) slowly expanding, cooler envelope, forming a planetary nebula. The term 'planetary' refers to the extended nature of the object. Therefore, the relatively slow wind that we described in the previous section—in which rather benign effects occur—has its relatively peaceful existence rudely disturbed. Firstly, there is the appearance of a harsh radiation field from the now much hotter central star. Instead of a weak infrared radiation field from the previously cool star, a powerful radiation field peaking in the ultraviolet will develop. Secondly, the new fast

Molecules in the Milky Way Galaxy

Table 4.4 Molecules detected in proto-planetary nebula CRL 618. Where necessary, linear (l) and circular (c) forms of a molecule are indicated.

Hydrocarbon species	C_2H, c-C_3H, l-C_3H, l-C_3H_2, c-C_3H_2, C_4H, C_5H, CH_3CCH
Species containing oxygen	CO, HCO^+, HOC^+, H_2CO, SiO
Species containing nitrogen	CN, HCN, HNC, C_3N, HC_3N, HC_5N, HC_7N, CH_3CN, CH_2CHCN, N_2H^+, $MgNC$
Species containing sulfur	CS
Large molecules	C_{60}, C_{70}, and PAHs containing both aliphatic and aromatic features

wind overtakes the original slower wind and drives a shock into it, raising the temperature of the shocked slower wind.

Therefore, when we observe a young planetary nebula we are looking at a region where there is a very wide range of physical conditions. While the hot fast stellar wind is molecule-free, the shocked gas where the two winds interact and the original envelope are sites where a rich chemistry can occur in potentially rather violent conditions (see Table **4.4**).

Clearly, the marked difference in the physical conditions, and their range, produces a significant difference in the chemistry of planetary nebulae compared to cool stellar envelopes. Methyl-substituted polyynes, ring structures such as benzene (C_6H_6), and fullerene structures such as C_{60} and C_{70} are present in young planetary nebulae though apparently absent in cool stellar envelopes. Some of these new species may arise in the degradation of carbonaceous grains; it's clear from changes in infrared spectra arising from the dust that significant changes occur in the dust, leading to the formation of ring structures rather than aliphatic. A young planetary nebula is a violent environment (see Figure **4.9**).

However, this chemically-rich phase is short-lived. After about a thousand years, the intense interaction between the radiation and the fast wind of the star with the original slow wind weakens, and so older planetary nebulae show less chemical variety. Eventually, all this circumstellar material will mix with the interstellar medium.

4.4.4 Chemistry in the Ejecta of Novae

The rate of evolution of a star depends rather sensitively on its mass. In a binary, *i.e.* two stars in orbit around each other, one star

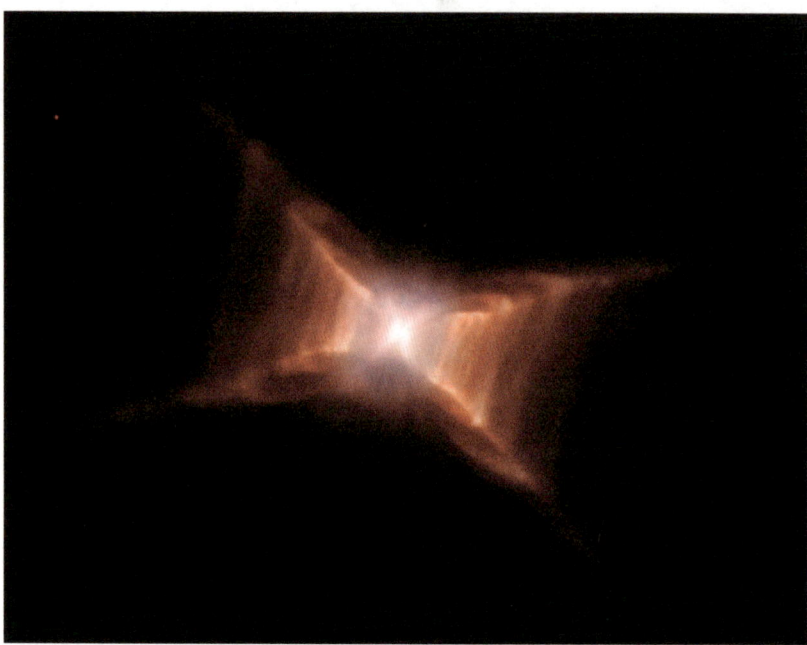

Figure 4.9 The Red Rectangle is a young planetary nebula. A thick torus confines the stellar outflow into conical shapes; we see these edgeon. Evidently, the outflow starts and stops, creating shocks, which generate structure in the image. Credit: NASA; ESA; Hans Van Winkel (Catholic University of Leuven), Martin Cohen (University of California, Berkeley).

may evolve much more quickly than its less massive partner. It is possible that the more massive star has reached the white dwarf stage (discussed in Section **4.4.3**) while the other still has an extended envelope. If so, then depending on the masses and on the separation between the stars, material from the extended envelope may overflow, *i.e.* be pulled gravitationally on to the white dwarf, which has a much stronger gravity at its surface than the other star. This transferred material is new fuel for nuclear fusion on the white dwarf—a star that has already used up all its fuel.

When sufficient fuel has accumulated on the surface of the white dwarf and been compressed by the white dwarf's intense gravitational field, a nuclear explosion occurs. We see this explosion as a nova, a sudden and large brightening of a star. About a dozen novae are detected each year. They are optically very bright indeed, sometimes as bright as one hundred thousand Suns. In the explosion, up to about 0.01 percent of a solar mass of material is ejected from the white dwarf with speeds of thousands of km per second. These ejecta are the ashes of the nuclear explosion, and are enriched in heavy elements (*i.e.* poor in hydrogen) compared to the solar relative abundances. The gas number densities in the ejecta are very high, comparable to those in stellar atmospheres (about 10^{12} H-atoms per cm^3).

Novae vary considerably in their properties, but many of them produce copious amounts of dust within two to three months after the initial outburst. We infer the presence of dust in a nova because the nova optical emission—initially very bright indeed—is suddenly almost extinguished while the infrared emission suddenly increases—a reliable signature of dust formation. Then, as the ejecta spread out the extinction caused by the dust is steadily reduced and the optical intensity partially recovers.

We really do not understand how dust is formed in a nova. But chemistry may give us a clue, for in the short period after the explosion and before dust appears, a rudimentary chemistry takes place and molecules are detected such as cyanogen (CN), molecular hydrogen (H_2), carbon monoxide (CO), silicon monoxide (SiO), silicon dioxide (SiO_2), silicon carbide (SiC), and some features that appear to arise in polycyclic aromatic hydrocarbons (see Section **4.1.2**). All these molecular detections are of emission bands arising from molecular vibration/rotation transitions.

Since novae that do not form dust show no molecular emission either, it suggests that chemistry is an essential pre-cursor to dust formation. Therefore, novae are really useful laboratories in which we can see—in real time—the results of microscopic processes that lead to chemistry and dust formation.

How can chemistry occur in a gas so very close to and irradiated by a hot white dwarf star? It seems—and is—a very hostile environment for chemistry. Of course, the density is very high, so collisions and reactions are very frequent, but the radiation field is very powerful and will dissociate and ionise molecules very quickly. In interstellar clouds, we rely on dust to reduce the intensity of the radiation field so that chemistry can succeed in forming molecules. But in novae ejecta the dust has not yet formed!

However, there is another kind of shielding that can help, due to atoms and molecules. For example, carbon atoms absorb radiation with wavelengths shorter than about 110 nm (in the far ultraviolet):

$$C + \text{radiation} \rightarrow C^+ + e$$

so carbon atoms can provide shielding for molecules that are dissociated by ultraviolet radiation with wavelengths shorter than this limit. This includes molecular hydrogen (H_2). In addition, molecular hydrogen itself is self-shielding (see Section **2.3.1**) because only very specific wavelengths are capable of dissociating H_2. In a thick cloud of H_2, radiation at these specific wavelengths is soon used up so that hydrogen molecules deeper into a cloud are protected by molecules nearer the edge of the cloud.

As long as carbon is mainly in the form of neutral atoms, then chemistry can occur, and—as in interstellar clouds—an efficient chemistry depends on the presence of molecular hydrogen (H_2). However, because we cannot invoke H_2 formation on dust because there is no dust! So other, slower, gas phase routes must be used (see Section **2.2.1**). These include three-body reactions:

$$H + H + H \rightarrow H_2 + H$$

when the density is high enough (above about 10^{11} H atoms per cm^3) and other routes dependent on electrons and protons at lower densities (as discussed in Section **2.2.1**). Molecular hydrogen can be collisionally dissociated (at high enough temperatures) in the reverse of the three-body reaction, and lost in photodissociation where the shielding is not complete.

The subsequent chemistry in the hot ejecta gas is then straightforward, for the high temperature means that many reactions that at low temperatures are inhibited by barriers or are slightly endothermic may proceed rapidly. For example, the reactions:

$$O + H_2 \rightarrow OH + H$$

$$C + H_2 \rightarrow CH + H$$

followed by:

$$C+OH \to CO+H$$

$$O+CH \to CO+H$$

form CO quite efficiently. This molecule is destroyed collisionally and by radiation. Diatomic carbon, which may have a role to play in forming carbonaceous dust, can be formed in similar exchange reactions in the hot gas:

$$C+CH \to C_2+H$$

$$C+CN \to C_2+N$$

$$C+CO \to C_2+O$$

All this chemistry proceeds only while the carbon in the ejecta is mainly in neutral atoms. Eventually, however, the radiation field from the white dwarf will eat its way through the ejecta and all the carbon will be ionised. At that point, chemistry ceases. If the role that chemistry has to play in forming dust has not occurred by this point, then dust will not form. It is unclear what that role is, but it may be similar to the formation of carbon chains in cool stellar envelopes (see Section **4.4.2**) though acetylenic ions or other similar routes. These and other carbon species might act as nucleation centres. If so, these carbon chain and other carbon species may then stick to these nuclei, forming carbonaceous solids on a timescale of less than one year. In this picture, dust formation in novae occurs by the accumulation into solids of carbon molecules that were formed in a rapid chemistry in the early phases of post-outburst evolution.

4.4.5 Chemistry in the Ejecta of Supernovae

The existence of molecules in the ejecta from a nova, *i.e.* a nuclear explosion at the surface of a hot star, seems unlikely enough. Yet—as we have seen—chemistry does find a way to proceed. However, a supernova explosion is not just a scaled-up nova, but a dramatically different enormous explosion and (apparently) an exceptionally hostile environment for chemistry to occur. It is remarkable that even in supernova ejecta chemistry does find a way to produce molecules in substantial abundances.

Supernovae come in various types. We shall discuss supernova explosions that occur when a massive star reaches the end of its life. On the basis of the number of remnants of supernovae that can be seen in the Milky Way, these events are believed to occur in the Milky Way on average about once every fifty years. However, the last actual detection of a supernova in the Milky Way was in 1604 (Kepler's supernova; see Figure **4.10** for an image of the remnant of this supernova). Many supernovae are detected in external galaxies.

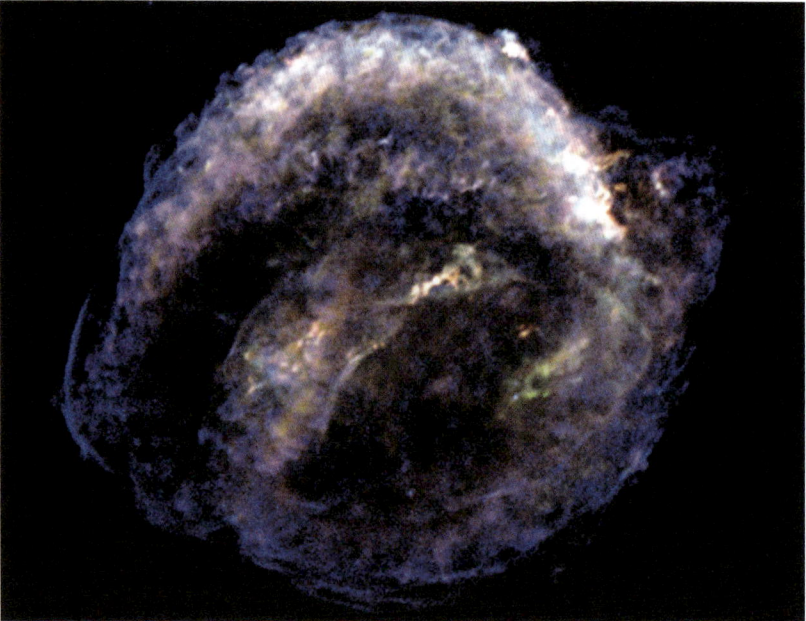

Figure 4.10 An X-ray image of the remnant of Kepler's supernova. Credit: NASA/CXC/SAO/D. Patnaude.

Supernova explosions are on an almost incredible scale. A single star becomes almost instantaneously as bright as an entire galaxy of many billions of stars, before fading from view in a few weeks or months. In the explosion, about a solar mass of material is ejected at enormous speed (up to tens of thousands of km per s) into the surrounding space, driving strong shocks into the interstellar medium. The post-shock gas is raised to a temperature on the order of 1 000 000 K, and radiates powerfully in X-rays. The total energy radiated by a star in the brief period of its supernova explosion and dissipation (only a few months) is comparable to the total energy radiated by the Sun in its entire lifetime of around ten billion years.

Yet even in this situation of super-intense radiation fields, hypersonic velocities and strong shocks, a high fraction of the ejecta can be molecular. CO and SiO are two examples of molecular species detected in the ejecta typically a few months after detection of the explosion. Dust formation may then follow the chemistry. For example, in 1987 a supernova (now called SN1987A) was detected in a neighbouring galaxy to the Milky Way, the Large Magellanic Cloud, some 170 thousand light years distant from the Sun. Molecular emissions were detected a few months after the SN1987A outburst was observed, and emission from warm dust was seen about a year after the outburst. The amount of dust has continued to grow; initially, the amount of dust was about 10^{-4} solar masses; by 2011 it had increased to about half a solar mass.

The chemistry in supernova ejecta is different from that discussed in any other interstellar or circumstellar context: it is *hydrogen-poor*; indeed, hydrogen may be entirely absent from some parts of the ejecta. Thus, in those H-free parts, all the chemical pathways that are normally invoked in interstellar and circumstellar chemistry—*i.e.* reactions with hydrogen molecules or with H_3^+—do not apply. Also, the chemistry precedes the detection of dust, so surface reactions cannot play a role. Clearly, this is unlike any other astrochemical environment. We will discuss first how a hydrogen-free situation can arise in normally H-dominated astronomy, and then describe a few formation routes for simple species in the ejecta of the supernova.

The main fuel in stars is hydrogen. Nuclear reactions in the interior of a star fuse hydrogen nuclei together to form helium nuclei, and the energy released in this contained and sustained 'hydrogen bomb' is the source of the star's radiation. But what happens when all the hydrogen that is at a sufficiently high temperature and pressure in the star's interior has been converted to helium? Without a heat source to maintain the pressure, the interior of the star begins to collapse and in so doing heats up again. For a sufficiently massive star, the temperature in the central regions can reach such a high value that a new set of nuclear reactions can occur: helium nuclei at the centre of the star begin to combine to form carbon nuclei, releasing yet more energy. At this stage, the star is developing a small carbon core with a helium shell, surrounded by unburned hydrogen in an outer layer.

The conversion of helium to carbon can continue in the central regions, where the pressure and temperature are high enough to drive this reaction, but eventually this reaction, too, ceases as the fuel, helium, is used up, and once again the star is without a central source of energy. If the star is sufficiently massive, further stages of collapse and heating drive a succession of nuclear reactions, ultimately ending with the formation of iron, the most stable nucleus, at the centre of the star. At this stage, the structure of the core of the star is of (i) iron in a central core, surrounded by distinct shells of mainly (ii) silicon and sulfur, (iii) oxygen and magnesium, (iv) oxygen, neon and magnesium, (v) carbon, (vi) helium, and all of these shells are surrounded by unburned hydrogen.

This, however, is the final stage for a massive star. Since iron is the most stable atomic nucleus, there is no other nuclear reaction that can rescue the situation by supplying energy to the star. Once all the silicon nuclei that are at a sufficiently high density and temperature have been converted to iron nuclei, then nuclear chemistry ceases. Then the star has no nuclear energy source, and it collapses under its own weight once again, this time to form a neutron star at its centre. All the enormous amount of gravitational energy that is suddenly released in the collapse drives the temperature high enough for some endothermic nuclear reactions to occur in which further element creation occurs, forming some nuclei more massive than iron. The collapse of star causes the chemically distinct shells to 'bounce off' the tiny neutron star at the centre: this is the supernova explosion. The shells are ejected in

Molecules in the Milky Way Galaxy

the supernova explosion, possibly undergoing some mixing. It is these inner ejecta that may be hydrogen-free.

How can CO, for example, be formed in these inner H-free ejecta? When CO was detected in SN1987A at about a few months after the outburst, the number density in the ejecta had already declined from values appropriate for stellar interiors to about a billion atoms per cm^3. At these densities, three body collisions are not effective. Instead, the chemistry of CO appears to depend on rather slow reactions, such as the radiative association:

$$C + O \rightarrow CO + \text{radiation}$$

which occurs perhaps only once in a million collisions. However, the number density is high so that enough collisions occur for this slow reaction to make a significant contribution to CO formation in the few months' time that is available. Other reactions may also contribute. For example, the formation of negative ions is favoured because—although much recombination has occurred—there is still some residual ionization remaining in the ejecta:

$$C + e \rightarrow C^- + \text{radiation}$$

followed by:

$$C^- + O \rightarrow CO + e$$

Similar schemes also work with O.

These production mechanisms for CO must compete with destructions by many routes in these hostile situations. Gamma rays (γ) from radioactive atoms such as the isotope of cobalt, ^{56}Co, created in the transient heating at the 'bounce', collide with and accelerate electrons to high speed. These electrons can dissociate molecules; they may also excite atoms to higher states that then relax to ground states emitting ultraviolet radiation that dissociates molecules.

CO appears in the material that was in the inner core of the pre-outburst star. The detected emission of protonated molecular hydrogen, H_3^+, in the supernova SN1987A obviously arose in the outer H-rich part of the pre-outburst star. To produce the implied large H_3^+ abundances we cannot rely on cosmic rays (as in dark clouds) to produce H_2^+, which is then converted to H_3^+ in reaction with H_2, as that process is far too slow. However, the γ rays from radioactive decay will certainly ionise molecular hydrogen:

$$H_2 + \gamma \rightarrow H_2^+ + e$$

and the H_2^+ then forms H_3^+ as usual, in reaction with another H_2 molecule.

There is also another route to H_2^+ that depends on electronically excited atomic hydrogen in a state with principal quantum number $n = 2$ or larger:

$$H(n=2) + H \rightarrow H_2^+ + e$$

The atomic hydrogen is excited in collisions, since the outer ejecta are quite hot, with a temperature of about 6000 K. In nearly all interstellar situations, these excited H atoms radiates so quickly that its population is very low. However, in the supernova ejecta, the density is high so the collision rate producing the excited atomic H is also unusually high.

4.5 CONCLUSIONS

The Milky Way has a remarkable variety of locations in which chemistry is occurring and where molecules are present at detectable levels. In fact, the ability of chemistry to form molecules in apparently quite hostile physical conditions is truly remarkable. Generally, since hydrogen is so much more abundant than other elements, the chemistry proceeds through reactions with hydrogen or its ions, until saturation is achieved. Then recombination with electrons or with other species generates new complexity. For chemists, this demonstration of the power of chemistry to occupy a wide variety of physical regimes is a delight. For astronomers, molecules are splendid tools. Astronomers use radiation from molecules to infer—often with the use of complex chemical kinetic models—the nature of the physical conditions where the molecules are located. This insight enables astronomers to determine the effects that molecules have in controlling by their radiation the ability of those regions to evolve.

Astronomers and chemists have worked constructively together to understand the processes by which the molecules are formed and destroyed in the rich variety of physical conditions, and this collaboration has led to new insights. However, the Milky Way is our own locality and—although there are many galaxies relatively nearby—we cannot expect to understand in similar detail the chemistry of very distant galaxies. Yet, the study of these galaxies may reveal information about galaxy formation and the origin of the Universe itself. The astronomical problem is that distant galaxies are spatially unresolved and that the molecular emissions will be dominated by the most powerful sources. In Chapter **6**, we shall consider how chemistry may differ in galaxies with physical conditions that vary from those in the Milky Way. The question we want to address there is this: can we use chemistry to determine the nature of unresolved galaxies? But first, we come even closer to home and look in the next chapter at the chemistry of planetary systems in the Milky Way galaxy.

FURTHER READING

T. W. Hartquist and D. A. Williams, eds, *The Molecular Astrophysics of Stars and Galaxies*, Oxford Science Publications, 1998.

J. Cernicharo and R. Bachiller, eds, *The Molecular Universe, IAU Symposium 280*, Cambridge University Press, Cambridge, 2011.

CHAPTER 5

The Path to Planets

5.1 ANGULAR MOMENTUM AND IONISATION

The planets of the Solar System lie in a plane because they formed in a spinning disk, much like the disks observed around many stars with masses similar to that of the Sun, but younger than the Sun with ages of roughly a million years or less. A spinning disk possesses angular momentum, and it is angular momentum that causes the existence of protoplanetary disks. Gravity leads to the collapse of material parallel to the axis about which such a disk orbits, while the centrifugal force associated with the angular momentum retards collapse towards that axis. Figure **5.1** shows an image of a disk obtained with the Hubble Space Telescope.

Consider a time before a disk has formed and imagine a spherical distribution of gas with a number density of hydrogen nuclei of about one per cubic centimetre, roughly the average for the interstellar medium. Suppose that this spherical cloud contains one solar mass of gas and, like the Sun, is at a distance of roughly 25 000 light years from the Galactic Centre. The galactic magnetic field constrains the cloud to keep one face towards the Galactic Centre as it follows an orbit, just as tides have caused the Moon to keep one face towards the Earth. Thus, the cloud rotates around an axis and therefore has angular momentum. For the cloud to collapse to a radius equal to the Sun's, it must lose all but roughly one ten thousandth of the angular momentum that it had initially.

The Cosmic-Chemical Bond
DA Williams and TW Hartquist
© DA Williams and TW Hartquist 2013
Published by the Royal Society of Chemistry, www.rsc.org

The Path to Planets

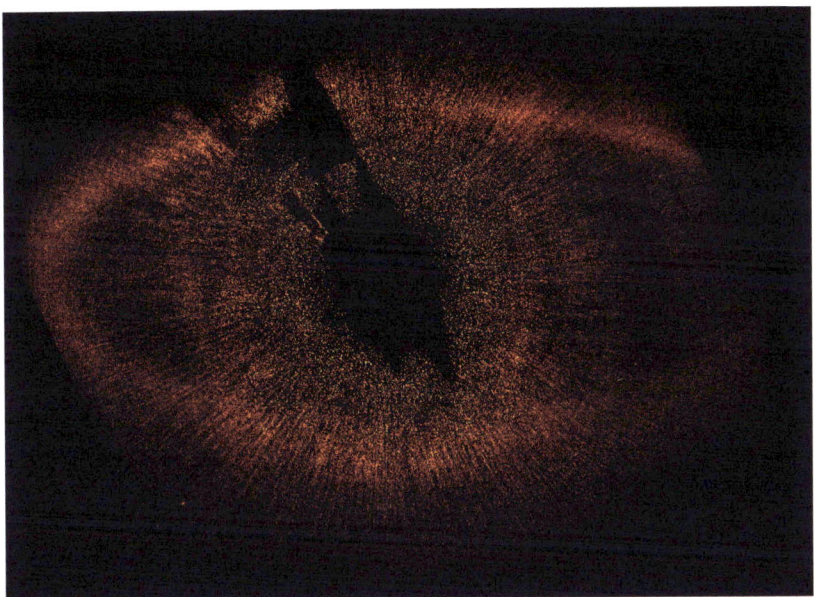

Figure 5.1 Hubble Space Telescope image of the dusty debris of a protoplanetary disk around the star Formalhaut. The disk contains the exoplanet Formalhaut B, which is not apparent in this image. (Credit: NASA, ESA, P. Kalas, J. Graham, E. Chiang, E. Kite (University of California, Berkeley), M. Clampin (NASA Goddard Space Flight Center), M. Fitzgerald (Lawrence Livermore National Laboratory), and K. Stapelfeldt and J. Krist (NASA Jet Propulsion Laboratory)).

As implied above, the magnetic field in a cloud couples its rotation to the motion of its surroundings. An ion or an electron is constrained by the field to orbit around the magnetic field lines with a frequency called the gyrofrequency. This frequency is proportional to the charge and the field strength and inversely proportional to the mass. Collisions of ions and electrons with neutral atoms and molecules mean that the entire gas is linked to the field lines. As a cloud collapses, this magnetic coupling transfers angular momentum out of the cloud. If too much angular momentum were lost, no disk and no planets would form. If too little angular momentum were extracted, the disk would be too extended and too tenuous for planets to form. Therefore, for planets to exist, the angular momentum transfer mechanism must turn off, and it must do so neither too early nor too late during the collapse.

It is the chemistry in the cloud that controls the angular momentum transport by determining the level of ionisation in the gas. When the number density of the ions drops to a low enough level the motion of the charged particles and the magnetic field decouple from the motion of the neutral gas. When this happens, the magnetic field ceases to transfer angular momentum efficiently. So an understanding of the chemistry establishing the ionisation structure of a collapsing cloud is vital for the construction of models of the dynamics of protoplanetary disk formation. In the next subsection we discuss how chemistry determines the level of ionisation in the cloud.

5.1.1 The Fractional Ionisation at the Onset of Disk Formation

At the current location of the Earth with respect to the Sun, the protoplanetary disk in which the Solar System formed once had a number density of up to 10^{14} hydrogen molecules cm^{-3}. For the moment, however, we will focus on dark star forming regions with densities that are a billion times less than that, but these lower density regions are the entities from which the disks are born. In the dark star-forming regions, the interaction of H_2 with cosmic rays having speeds of tenths of that of light, induce most of the ionisation. About 97% of the ionisations create H_2^+, which reacts rapidly with H_2 to form H_3^+—protonated molecular hydrogen— (see Section **2.3.2**). In the dark star-forming regions in which the depletion of CO onto dust grains is modest, most of the H_3^+ is removed in reactions with CO:

$$H_3^+ + CO \rightarrow HCO^+ + H_2$$

When CO is not too highly depleted, most ionisations lead to the formation of the formyl radical ion HCO$^+$, which is usually the most abundant molecular ion. The importance of H_3^+ removal by dissociative recombination:

$$H_3^+ + e \rightarrow H_2 + H$$

increases with increasing depletion of CO and other gas phase species.

Millimetre wave radiation emitted by HCO^+ in interstellar clouds was first reported in 1970, but the observed features were initially unidentified. The Canadian Nobel Prize winning spectroscopist Gerhard Herzberg, who made important contributions to astronomy, was amongst those suggesting that the newly detected features were due to HNC. However, in the same year the Harvard physical chemist William Klemperer correctly suggested that the emission is due to rotational transitions in HCO^+. Additional support for this identification was published in 1973 in a paper in which he and Eric Herbst argued that a cosmic-ray triggered interstellar chemistry would lead to HCO^+ being an abundant ion. Subsequently, quantum calculations of the potential surfaces of HCO^+ and other molecules confirmed that HCO^+ emission had been detected.

In a dark star-forming region, the two most significant processes for the removal of HCO^+ are dissociative recombination:

$$HCO^+ + e \rightarrow CO + H$$

and charge transfer with gas phase metallic atoms:

$$HCO^+ + M \rightarrow HCO + M^+$$

where M is used to designate any metallic element, *e.g.* Na or Mg. Usually dissociative recombination will dominate. In 1974 the Harvard scattering theorist, atmospheric scientist and astrophysicist Alex Dalgarno and Michael Oppenheimer (who subsequently became an influential environmental scientist) pointed out that the charge transfer reaction is also important because metallic atomic ions are the most abundant positively charged species in star forming regions. The high abundance of metallic ions compared to molecular ions is due to the main gas phase removal mechanism of such atomic ions being radiative recombination, a reaction that, typically, is at least several orders of magnitude slower than dissociative recombination.

Metal ions, M^+, are removed primarily in collisions with dust grains. In a dark region with a number density of about 10^5 hydrogen molecules cm^{-3}, most dust grains carry a single negative elementary charge, a consequence of the electrons having higher thermal speeds than the ions and colliding more frequently with grains. So ion-grain collisions lead to neutralisation.

For a cosmic ray ionisation rate of 10^{-17} s^{-1} and reasonable assumptions about the grain population and depletions, the fractional ionisation (relative to hydrogen) is roughly 10^{-8} at a number density of 10^5 hydrogen molecules cm^{-3}. This fraction scales inversely with the square root of the number density.

5.1.2 Inference of the Fractional Ionisation and Deuterium Fractionation

In Section **4.3.3** we described how ion-molecule reactions at sufficiently low temperature can tend to concentrate deuterium (D) in preference to hydrogen (H) in some molecules, a process known as fractionation. Some efforts to estimate the fractional ionisation in dark star-forming regions from observations have involved considerations of deuterium fractionation. The cosmic abundance ratio of deuterium nuclei to hydrogen nuclei is about 10^{-5}, and the HD to H_2 abundance ratio in star-forming regions is near the cosmic ratio. However, in many of the same regions the ratios of deuterated to protonated versions of some molecular species are several orders of magnitude larger than 10^{-5}. For example, ND_3 is a trebly deuterated species of ammonia (NH_3) seen in some star forming regions in which the gas phase depletions of molecules, including CO, are substantial.

As we have seen in Section **4.3.3**, a key deuterated species is H_2D^+. In a star-forming region at a temperature of 10 K, the equilibrium ratio of the abundances of H_2D^+ and H_3^+ would be over a billion times the ratio of the abundances of HD and H_2 if the species were not removed by any mechanisms other than those two reactions. In fact, H_2D^+, like H_3^+, is removed in reactions with species like CO and by dissociative recombination. In many regions those reactions bring the equilibrium ratio of the abundances of H_2D^+ and H_3^+ down into the range of about 100 to 1000 times the HD to H_2 abundance ratio. The largest deuterium fractionation occurs in those cold regions in which gas phase species containing elements more massive than helium are highly depleted by incorporation into ices on dust grains, and the electron density is also particularly low.

The significant contribution of dissociative recombination to the removal of H_2D^+ implies that, in principle, the abundance of electrons can be inferred from the ratio of the H_2D^+ and H_3^+ abundances. To make the inference, the abundances of other species, including CO in particular, that remove H_2D^+ must be known. The ratio of the H_2D^+ and H_3^+ abundances is commonly deduced from measurements of the ratios of abundances of the deuterated and protonated versions of other species. For example, the ratio of the DCO^+ and HCO^+ abundances has been used. In the simplest model, the abundance ratio of DCO^+ and HCO^+ is assumed to be one third that of H_2D^+ and H_3^+. This assumption is made because one might expect DCO^+ and HCO^+ to be removed on the same timescales and a third of the reactions of H_2D^+ with CO to lead to DCO^+ and two thirds to form HCO^+. N_2D^+ and N_2H^+ are also observed in efforts to use deuterium fractionation to estimate the fractional ionisation.

Studies of deuterium fractionation suggest fractional ionisations in the dark star-forming regions in the range from about 10^{-8} to 10^{-6}. The higher fractional ionisations may be due to locally enhanced ionisation rates and also to smaller than usually assumed ratios of grain cross section to gas phase neutral particle number density. Star-forming regions often contain stars with powerful winds, which can drive shocks in which cosmic rays are accelerated and gas is heated to high enough temperatures to lead to the emission of ionising X-rays. Grain–grain collisions in dense regions may be frequent enough for coagulation to make more large grains so that the total surface area of grains is significantly reduced.

5.1.3 'Small' Grains, Anions, and Ionisation

The size distribution of grains is important for the fractional ionisation, because the grain surfaces are the sites where metallic atomic ions recombine with electrons.

The size distribution of interstellar grains has been investigated with observations of how optically thin interstellar clouds affect the light from stars behind them. These show that there are many more small grains than large (see Section **1.2.1**). In such a cloud, the total surface area of grains with sizes of 10 to 20 nm is a factor of about three greater than the total surface area of grains with sizes of 100 to 200 nm. The largest grains in the optically thin clouds are probably smaller than a micron, but the lower limit on the size of grains in them is not certain. As mentioned above, coagulation may lead to significant modifications of the grain size distribution, but other processes, like sputtering in shocks, may do so as well. A value of about 10^{-8} for the model fractional ionisation is quoted in Section **5.1.2**, but this result is based on the assumption that few grains are much smaller than 100 nm.

In contrast, some researchers have supposed that the grain size distribution extends below 10 nm and down to the scale of individual polycyclic aromatic hydrocarbon (PAH) particles. If so, the total recombination cross-section of 'small' grains or 'big' molecules may be much larger than that of grains with sizes of about 100 nm. The 'small' grains would be sinks for electrons, and one can imagine parameters and conditions that would result in the most abundant negatively charged particles in an interstellar cloud being 'small' grains. If they were, recombination of metallic ions with them would lead to a reduction in the fractional ionisation.

The preceding paragraph refers to 'small' grains or 'big' molecules. The distinction between the two is blurred. Also many astrochemists are rather more easily impressed by the size of a molecule than most other chemists. Some astrochemists consider formaldehyde a big molecule. So a molecule like C_4H, not to mention C_6H or C_8H, may seem big to some. However, such molecules, unlike 'small' grains, cannot be major sinks of electrons in interstellar clouds. Still these molecules do have sufficiently large electron radiative attachment rate coefficients that emissions, arising in star-forming regions, of their anions, C_4H^-, C_6H^- and C_8H^-, have been detected following dedicated laboratory studies of their spectra.

5.1.4 Ionisation at Higher Densities and in Disks

The drop in fractional ionisation with increasing density implies that above a particular hydrogen number density, the number density of gas phase electrons falls below the number density of dust grains. That particular hydrogen number density depends on several parameters, including the ionisation rate, and the grain population, but is around 10^9 hydrogen molecules cm^{-3} for some reasonable assumptions. At higher hydrogen number densities, dust grains become the dominant carriers of negative charge. At hydrogen number densities exceeding about 10^{11} cm^{-3} to 10^{12} cm^{-3}, dust grains also become the dominant carriers of positive charge. The nature of hydromagnetic phenomena, including those affecting angular momentum transport, changes when the grains become important carriers of charge and current. The change is due in part to charged grains having much lower gyrofrequencies around the magnetic field than electrons and gas phase ions.

The ionisation rate throughout much of a protoplanetary disk is probably lower than in a substantially more tenuous star-forming region because the column density of material in the disk can shield the interior of the disk from the cosmic rays. In addition, young stars with disks have magnetised outflows that may partially exclude cosmic rays. A lower bound to the ionisation rate in a disk is provided by the radioactive decay of some unstable species, including the isotopes ^{26}Al and ^{40}K. Of course, the reduction in the ionisation rate affects the magnetohydrodynamic processes leading to angular moment transport.

A key question concerns the locations and extents of so-called 'dead zones' in protoplanetary disks. Such zones are sufficiently weakly ionised that a process referred to as the magneto-rotational instability, or MRI, is ineffective in generating turbulence, which plays a major role in angular momentum transport. The existence of such zones affects the accretion rate and final mass to which a star will grow and also how rapidly young planets migrate due to interactions with the disk. So the chemistry controlling the ionisation also determines the properties of the planetary system forming in a disk.

The outer edge of a disk is exposed to radiation from the central star and any nearby stars, and it may interact with an outflow from the central star. These effects maintain a much higher fractional ionisation and higher temperature in the outer layers than those of the disk regions considered above. The fractional ionisation and the temperature drop towards the central mid-plane. The fractional ionisation in the cooler regions can be studied with interferometric millimetre and submillimetre observations of species mentioned in Section **5.1.2**. These include H_2D^+, HCO^+ and N_2H^+ (protonated molecular nitrogen). An early attempt to infer the fractional ionisation in regions of the DM Tau protoplanetary disk indicate that it is in the range of about 3×10^{-11} to about 4×10^{-10} where the temperature is 20 K and below.

5.2 VERY BASIC NEUTRAL CHEMISTRY DURING COLLAPSE AND DISK FORMATION

In the dense core from which a solar-like star and a disk form, most of the molecules containing elements more massive than helium will be frozen onto the surfaces of grains. H_2O and CO ices are the most abundant surface species, but O_2, CH_4, NH_3, N_2 and NO are also major constituents of the mantles. As the core collapses some heating, due mostly to the radiation of the young central star, of material destined for the disk occurs. The thermal history of any particular parcel of material depends on how far from the star it is when it enters the disk and on when it enters the disk. So the chemistry is spatially and temporally dependent.

The thermal history also depends on the mass of the disk formed, which depends on properties of the dense core. The qualitative description of the chemistry given throughout the Sections **5.2.1**–**5.2.5** is based on results, published in 2011 by Ruud Visser, Steven Doty and Ewine van Dishoeck, for a disk containing about 0.13 solar masses and extending out to about 47 Astronomical Units (AU). Recall that an AU is the distance of the Earth from the Sun. The Solar System's outermost planet, Neptune, is about 30 AU from the Sun, and the orbit of Pluto, now no longer classified as a planet, extends from about 30 AU to about 50 AU from the Sun. The half-thickness of the disk increases roughly linearly, from zero to about 6 AU, with radius as the radius increases from zero to about 15 AU. It is around 6 AU to 7 AU out to a radius of about 40 AU, beyond which it decreases.

5.2.1 Carbon

In the dense core most of the carbon is in CO ice, but as the gas falls closer to the disk the temperature increases. The increase causes any CO that is not trapped in cavities within the H_2O ice to evaporate. This occurs well before the gas reaches the disk. Throughout the disk the temperature is above 20 K, which is more than sufficient to prevent any of the evaporated CO from returning to the solid phase.

Surface CH_4 that is not trapped in the H_2O ice also evaporates but somewhat later during the infall than the corresponding CO. As described in the next subsection, photodissociation is an important mechanism for the removal of some gas phase molecules throughout a small fraction of the disk's mass, particularly near the inner edge of the disk. CH_4 is dissociated by radiation with wavelengths of 145.0 nm and less in such regions. CO is removed by photons with wavelengths of 107.6 nm and less, and the central star is too cool to produce sufficient radiation at these shorter wavelengths for photodissociation to affect CO much even throughout most of the inner edge of the disk.

CO is removed by only a few reactions. Reactions with H_3^+ form HCO^+, but dissociative recombination forms CO again. He^+, produced by cosmic ray induced ionisation of He, destroys CO on a timescale that affects the CO abundance little. However, it does produce C^+, which reacts to initiate sequences that have non-negligible effects on other species, *e.g.* CH_4.

5.2.2 Oxygen

In the dense core most of the surface oxygen that is not contained in CO is in H_2O ice, and most of the H_2O that is on the surface remains there, at least for a good fraction of the lifetime of the disk. H_2O is significantly more strongly bound to the surface than any CO that is not trapped within the water ice, and temperatures enough in excess of 100 K to allow the H_2O to evaporate obtain in only about ten percent of the disk material. That ten percent lies within about 10 AU of the star. The H_2O ice survives in all material located in the mid-plane of the disk further than 6 AU from the star. In a few percent of the inner material H_2O and CO are the most abundant gas phase species. Closer to the star gas phase atomic oxygen, O, is abundant due to the photodissociation of H_2O and OH. All of the gas phase atomic oxygen is within 10 AU of the star and within 1 AU of the surface of the disk.

Some material warms as it falls towards the disk, cools as angular momentum transport in the disk induces it to move outwardly and then warms as it moves inwardly again. Consequently, roughly a few percent of the H_2O evaporates, then freezes-out and then evaporates again.

Like CO, O_2 is in the gas phase throughout almost all of the disk material, though it evaporates somewhat later than CO during the infall. Photodissociation of O_2 plays some role in the disk chemistry.

5.2.3 Nitrogen

In the dense core, N_2 ice contains about 80% of the nitrogen. N_2 has a similar binding energy to that of CO and, like CO, reacts with few species and is not removed by photodissociation. Consequently, its distribution in the disk is similar to that of CO.

The NH_3 and NO ice abundances are comparable.

NH_3 ice in the disk evaporates at about 73 K, a temperature roughly five times that at which CO evaporates and somewhat less than a factor of two lower than that at which H_2O evaporates. NH_3 is photodissociated in the same regions in which H_2O is photodissociated. Throughout about a third of the mass of the disk, NH_3 formed prior to the mass accumulated in the disk is contained in ice. About three quarters of that NH_3 evaporated during the infall stage and froze-out again once the material containing it was fully incorporated in the disk. In much of the remaining disk material, NH_3 evaporated during infall and was photodissociated or destroyed by HCO^+. In the fraction of such material that ends up in dark enough parts of the disk, some NH_3 is re-formed by the gas-phase reactions that produce it in molecular clouds, but the abundance of NH_3 remains below that of NH_3 ice in the precursor core.

The abundance of the NO ice shows a pattern similar to those of the CO and N_2 ice abundances. Unlike gas-phase CO and N_2, gas-phase NO is removed by photodissociation.

5.2.4 Abundances Changed by Subsequent Disk Chemistry

Clearly the temperatures in some regions of a disk are sufficiently high to maintain most species at reasonable abundances in the gas phase. This allows chemistry occurring in the disk to modify the abundances created during the formation of the disk. Some of the species with abundances that are significantly changed by later disk chemistry are: gaseous O, N, O_2, OH, CH_4, NH_3 and NO; NH_3 ice. However, the abundances of some species change little subsequent to disk formation. These species include: H_2O gas and ice; O_2 ice;

CO gas and ice; CH_4 ice; C; C^+; N_2 gas and ice; NO ice; HCO^+ and N_2H^+.

5.2.5 'Complex' Organics?

The molecules treated earlier in this section are simple, but they are the precursors of more complicated species. Small first-generation organic molecules, like CH_3OH, are formed on dust in dense cores prior to collapse. CO must be present on the grain to produce CH_3OH and other precursors of some moderately larger species. However, once those precursors become abundant in the ice, CO does not need to be present for the production of the larger species. This is important because the icy reactants that lead to the larger species are not very mobile on the surfaces when the temperature is low enough to retain nearly pure CO ice. The reactants have higher binding energies than CO and do not evaporate at temperatures of 20–40 K (see also Sections **3.4.3** and **4.3.2**). The formation of some larger organics, including methyl formate ($HCOOCH_3$), can occur on grains if they spend several tens of thousands of years at temperatures of 20–40 K. Much of the material going into the disk spends sufficient time in this temperature range during the infall phase that the surface formation of organics larger than CH_3OH is likely to occur.

5.3 CHEMISTRY IN MORE MASSIVE DISKS

Most of the disks surrounding solar-like stars that have been discovered so far have masses that do not exceed the mass of the disk considered in the previous section (about one tenth of a solar mass). However, a few young stars behave like FU Orionis, which was observed in 1937 to brighten by a factor of several hundred over a few months. The cause of this brightening is not firmly established, but some have argued that it is associated with instabilities occurring in a disk that is massive enough for its self-gravity, due to the material that it contains, to drive dynamics that the gravity of the central star alone would not. A disk containing as much as 0.4 to 0.5 solar masses might exist around a solar-like star. Indeed, such massive disks may surround most very young solar-like stars. If so, they can survive for only tens of thousands of years, lose mass and become more like the commonly observed disks around young solar-like stars. One of those more common disks survives roughly a million years.

If the self-gravity in a disk is sufficiently strong relative to the gravity of the central star, the disk is gravitationally unstable and spiral structures result. They look similar to the spiral arms seen in some galaxies. Material that moves through them is heated by shocks and lifted away from the central plane of the disk. The lifting results in some material that was previously screened by other material from the stellar radiation field becoming exposed to direct illumination. Depending on how far the shock occurs from the central star, the exposure can lead to the photodissociation of some species.

In 2011 John Ilee, who was then a postgraduate student at the University of Leeds, Aaron Boley of the University of Florida and their collaborators simulated the chemical composition of a gravitationally unstable protoplanetary disk. They restricted consideration to material that is sufficiently far into the disk and, thus, away from the disk surface that photodissociation could be neglected.

5.3.1 Properties of the Disk

The gravitationally unstable disk studied by Ilee, Boley and their colleagues contains 0.39 solar masses. The outer radius of it is about 60 AU. The temperature throughout the entire disk is sufficient for CO ice to evaporate, but throughout the disk most of the H_2O is contained in ice. However, the temperature is higher in the spiral features and H_2O ice evaporates in a substantial fraction of the material contained in them.

5.3.2 Species Affected Primarily by Desorption and Adsorption Only

As mentioned earlier, the lifetime of a disk in which gravitational instability is important is only tens of thousands of years. As noted in Section **5.1.4**, the ionisation rate in a disk is likely to be smaller by several orders of magnitude than in dense cores. The abbreviated life of a disk and the low ionisation rate contribute to some species in dark regions of a massive disk being affected

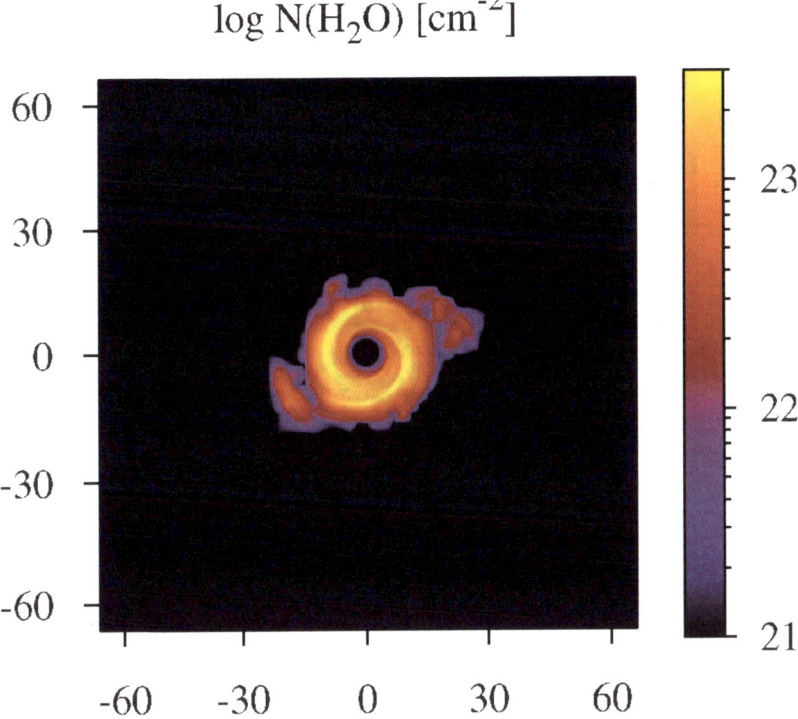

Figure 5.2 Column density of H_2O in a model disk in which self-gravity is important. Adapted from Figure 7 of Ilee et al. (2011).

very little by gas phase chemistry. The abundances of such species are affected in a gravitationally unstable disk primarily by desorption resulting from shock heating in the spiral features and adsorption subsequent to cooling. H_2O, NH_3 and H_2CO are amongst such species. Figures **5.2** shows the model column density distribution of gas phase H_2O.

5.3.3 Species Also Affected by Gas-Phase Reactions

Gas-phase chemistry does affect the abundances of some species in dark regions. For example, H_2O rapidly removes HCO^+, and in regions where shock heating returns most of the H_2O to the gas phase, HCO^+ is notable by its scarcity. Figure **5.6** shows the model column density distribution of gas phase HCO^+. Gas phase chemistry also affects a number of species formed primarily by neutral−neutral

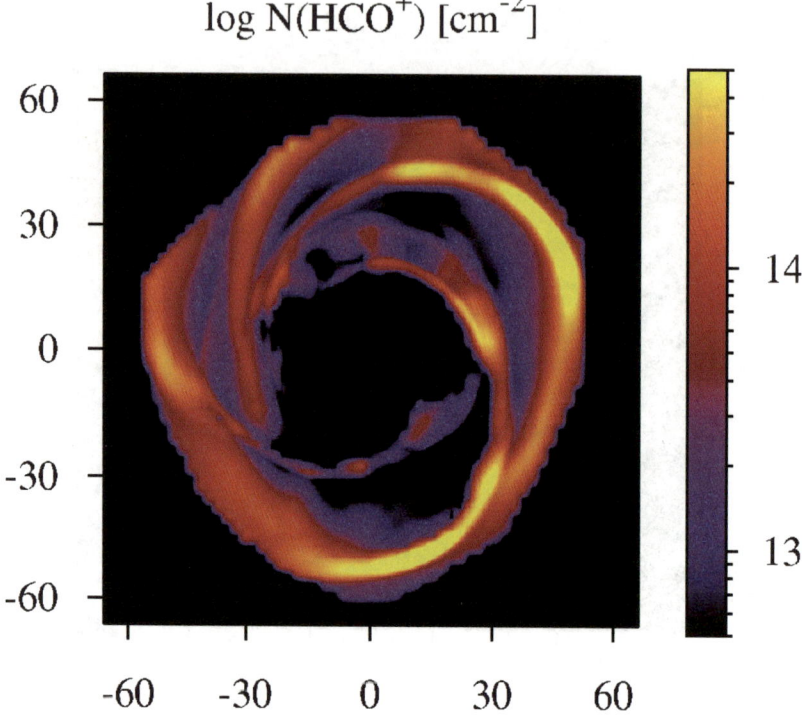

Figure 5.3 Column density of HCO^+ in a model disk in which self-gravity is important. Adapted from Figure 8 of Ilee *et al.* (2011).

reactions. These include some sulfur-bearing species (*e.g.* CS and SO_2) and some nitrogen-bearing species (*e.g.* OCN and HNO).

5.4 CHEMISTRY IN MORE EVOLVED DISKS

After a protoplanetary disk has formed and evolved through its most massive phase, it will survive for roughly a million years. During most of its life, it will have a mass of about 0.1 solar mass or less. The contribution of such a protoplanetary disk to the gravitational field is too small, compared to the contribution of a central solar-like star, for spiral dynamical features to arise due to the disk's self-gravity. Most models of the chemistry in protoplanetary disks have been developed for such evolved disks. The models are usually based on the assumption that a disk is

The Path to Planets 133

axisymmetric and has a density scale-height at any given distance from the central star that is small compared to that distance.

5.4.1 Typical Disk Properties

Catherine Walsh, Hideko Nomura, Thomas Millar and Yuri Aikawa are amongst the researchers to have constructed axisymmetric models of the chemistry of evolved protoplanetary disks in which the self-gravity is unimportant. Their work yielded the results addressed in this and the following subsection. Figures **5.4** and **5.5** show the number density and temperature distributions for one of the physical models used in their chemical studies. The heating mechanisms include viscous dissipation and the injection of photoelectrons into the gas phase, following the absorption of the central star's far ultraviolet and X-ray radiation and interstellar background far ultraviolet radiation. Other stars provide the sources of the interstellar background radiation. Dust grains absorb photons and emit infrared and submillimetre continuum radiation, some of which is absorbed in the disk. Collisions between gas and

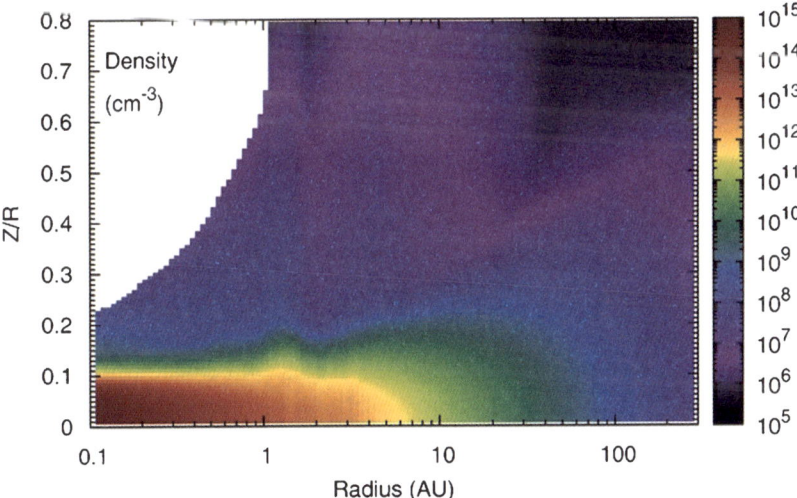

Figure 5.4 The number density in an axisymmetric model of an evolved disk. The radius is the distance from the symmetry axis, and z/R is the ratio of the height above the mid-plane to the radius. Adapted from Figure 2 of Walsh *et al.* (2012).

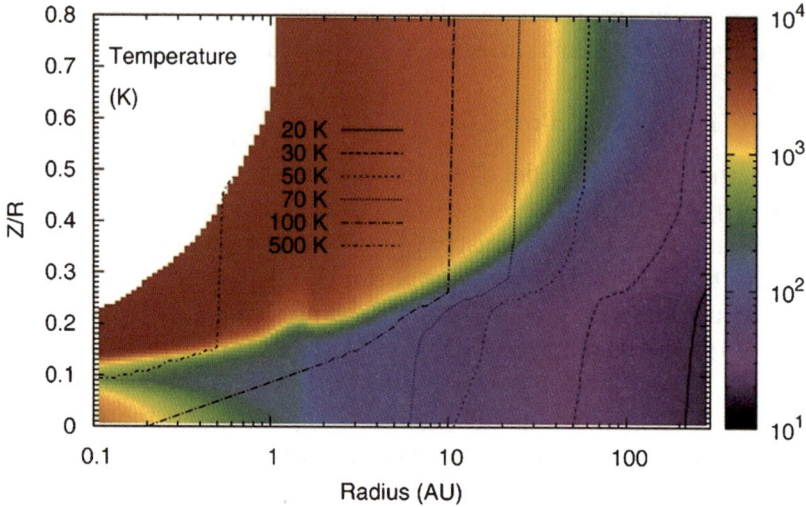

Figure 5.5 The temperature in an axisymmetric model of an evolved disk. The radius is the distance from the symmetry axis, and z/R is the ratio of the height above the mid-plane to the radius. Adapted from Figure 2 of Walsh *et al.* (2012).

dust transfer energy. The continuum radiation of the dust and line radiation of molecules carry energy away.

5.4.2 Molecular Distributions

Figures **5.6** and **5.7** show the gas phase fractional abundances of H_2O and HCO^+, respectively, in one of the models developed by Walsh and collaborators. The smallest values occur at the parts of the edge of the disk nearest to the star and at the part of the mid-plane of the disk that is not near to the central star. Photodissociation and photoionisation are responsible for the small gas phase fractional abundances near the inner edge of the disk. Freeze-out of molecules onto grains is responsible for the small gas phase fractional abundances occurring throughout most of the mid-plane.

The highest fractional abundances of gas-phase species occur in regions shielded sufficiently by material near the disk edge for photoionisation and photodissociation to be reduced substantially. However, in the regions in which the highest fractional abundances of gas phase species occur, enough heating, due to photoabsorption, takes place for molecules to remain unbound to the grains.

The Path to Planets

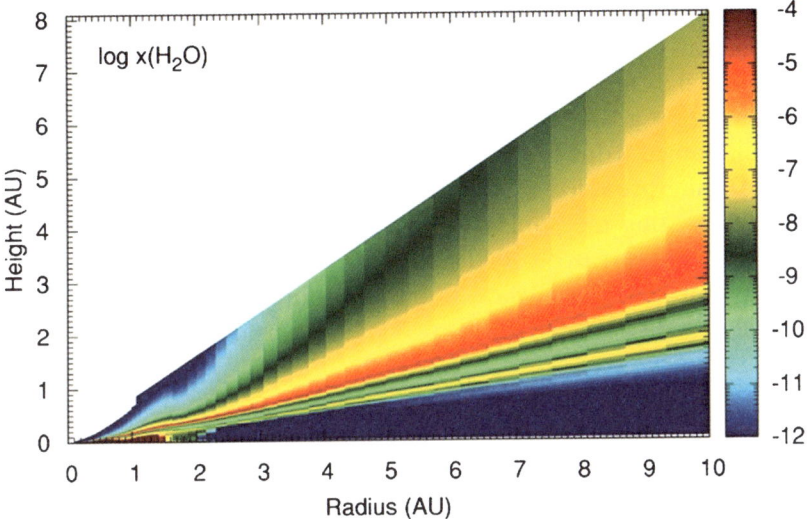

Figure 5.6 The fractional abundance of H_2O in an axisymmetric model of an evolved disk. The distance from the symmetry axis and the height above the mid-plane are measured in AU. Adapted from Figure 3 of Walsh *et al.* (2010).

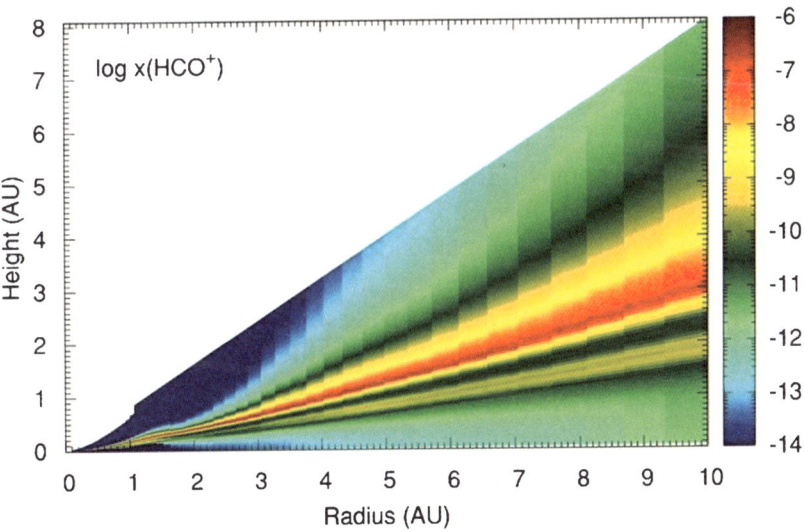

Figure 5.7 The fractional abundance of HCO^+ in an axisymmetric model of an evolved disk. The distance from the symmetry axis and the height above the mid-plane are measured in AU. Adapted from Figure 2 of Walsh *et al.* (2010).

The relationship between the column densities of H_2O and HCO^+ for lines-of-sight perpendicular to the mid-plane for an axisymmetric disk model differs markedly from the corresponding relationship for the model of a disk in which self-gravity is important. Ultimately, this difference is due to the heat required to keep molecules in the gas phase being provided by different mechanisms. Photoabsorption is the key heating mechanism in regions where gas phase species are abundant in the evolved disk model. Dissipation of mechanical energy, particularly in shocks, plays a more important role in the model of a disk in which self-gravity is important.

5.5 COMETS

In the Solar System, comets probably formed before the planets. Thus, the abundances and properties of atomic and molecular species that comets contain may provide information about the early phases of the Solar System.

Traditionally comets have been classified by the properties of their orbital dynamics. The two main classes are described as being 'nearly isotropic' and 'ecliptic'. The nearly isotropic comets are associated with the 'Oort Cloud' at tens of thousands of AU from the Sun. The ecliptic comets come from the 'Kuiper Belt', which extends from about 30 AU to 100 AU from the Sun. Some evidence of a third significant repository of comets exists.

Though the Oort Cloud and the Kuiper Belt are distinct, substantial fractions of the bodies in both may have common origins. These bodies formed nearer to the Sun than they are now; scattering with the giant planets caused them to move further from the Sun. Consequently, the abundances and properties of chemicals in the comets may provide an even more fundamental means of classifying the comets than the present dynamics do. Efforts to develop a taxonomy based on the chemicals in the comets are ongoing.

5.5.1 Measuring the Inventory of Cometary Volatiles

Most of the material of a comet is composed of mixtures of refractory solids and ices. When a comet is near enough to the Sun for sufficient heating of the ices to occur, solid refractory particles and gases are released. In the vicinity of the comet head, the relative abundances of different gaseous molecular species are mostly the same as those that characterized the ice that has melted. However, the relative abundances may be affected by the injection of species from the non-icy components of the dust. Further from the head, the solar radiation photoionises and photodissociates the primary evaporated species and creates product species.

Product, as well as primary, species radiate. For example, optical emission of OH arises due to fluorescence. Also, OH infrared emission results from the photodissociation of H_2O producing excited OH. CN, C_2, C_3 and NH are amongst the other molecular species detected in the optical emission of comets. Neutral atomic oxygen is also observed in optical emission, following its production by the photodissociation of OH. H_2O, CO, H_2CO, CH_3OH, CH_4, C_2H_2, C_2H_6, HCN, NH_3 and OCS are some of the species detected in cometary infrared emissions. The infrared emissions of some of these species cannot be observed from the ground, but data obtained with space-based infrared telescopes have become increasingly available.

Radio and millimetre emissions from a number of cometary gaseous species are also detectable. Such species include CO, H_2CO, CH_3OH, HCOOH, HCN, HNCO, CH_3CN, HC_3N, NH_3 and H_2S.

Remarkable space missions have led to encounters between probes and a few comets. Samples from several have been delivered to Earth, leading to a number of interesting results. One is the discovery of evidence of glycine (NH_2CH_2COOH) in the material from comet Wild 2. NH_2CH_2COOH is the simplest of amino acids, which are addressed further in Section **5.6.2**.

However, most information on the compositions of comets has been provided by remote optical, infrared, radio and millimetre observations. The focus of the remainder of Section **5.5** is primarily on results obtained from optical and infrared observations.

5.5.2 Compositional Classification Based on Optical Data

H_2O is the most abundant gas in a cometary outflow, as measured by observations above the Earth's atmosphere. Often the abundances of other gaseous species in such an outflow are measured relative to the abundance of H_2O, or the abundance of a species used as a proxy for H_2O. Such proxies are employed because H_2O is not observable with ground-based optical and infrared telescopes. They include OH, which is a product of H_2O photodissociation, and HCN, though in cases in which both H_2O and HCN data are available, some differences, from comet to comet, in the measured values of the HCN to H_2O abundance ratio are noted.

In 1995 Michael A'Hearn, of the University of Maryland, and his colleagues reported an important attempt to establish cometary taxonomy based on molecular composition inferred from optical data. They found that the C_2 to CN abundance ratios have very similar values for many comets but are lower by factors of about three to somewhat more than ten for others. The C_2 and CN molecules are product molecules created by photodissociation, and their abundance ratio is supposed to reflect the ratio of the abundances of molecules having carbon chains and those that do not.

The comets displaying deficiencies in carbon−chain species belong to the sub-class of ecliptic comets having orbits that are influenced significantly by Jupiter's gravity. In contrast, radio and millimetre-wave measurements of abundance ratios show no obvious trends associated more with this sub-class of comets than with the nearly isotropic comets.

Further substantial work involving the consideration of optical data has led to the identification of as many as ten compositional cometary classes. Several of these are sub-classes of the two initially identified by A'Hearn and his collaborators.

5.5.3 Compositional Classification Based on Infrared Data

Due to CO having a smaller dipole moment than most molecules and CH_4 having no dipole moment, CO and CH_4 desorption occurs more readily than the desorption of many other species. Infrared observations reveal that the CO and CH_4 abundances relative to H_2O vary from comet to comet by at least an order of magnitude but show no clear correlation with one another.

C_2H_6, CH_3OH, HCN and C_2H_2 are believed to be primary species, and infrared data show that their abundance ratios relative to H_2O vary in a correlated manner from comet to comet. The abundances of these gaseous organic species relative to H_2O have been used to identify three compositional classes.

The Path to Planets 139

For the first class of comets these abundances vary little from comet to comet. Each member of this class has an orbit that implies that it is an Oort Cloud object.

The second class seems to be less prevalent. Some of its members have split or experienced major outbursts. They all have higher abundances of C_2H_6, CH_3OH, HCN and C_2H_2 relative to H_2O than any of the comets in the first class. The comets in the second class include bodies from both the Oort Cloud and Kuiper Belt.

The comets in the third class have abundances of the four key species that are small compared to the abundances in the set of comets belonging to the first class. A comet belonging to the third class can be either an Oort Cloud object or be associated with the Kuiper Belt.

5.5.4 Comparison with Interstellar Chemistry

As described in Section 5.2, during the accretion of dense core material onto a disk, the chemical composition of some of the dense core material is altered by photodissociation. However, much of the material that falls onto the disk is shielded from or far enough from the central star for photodissociation to play much of a role. If comets form from such material, in which photodissociation was not important during the accretion phase, one might expect the chemical composition of at least some comets to resemble that of a star-forming region where the number density of molecular hydrogen is in the range of 10^4 to 10^8 cm^{-3}. Comparisons have been made of the abundance ratios, with respect to water, of gaseous species near the heads of comets and the corresponding abundance ratios of species in star forming regions. They indicate that the typical composition of cometary ice is consistent with that expected if the ice were formed and processed in star forming regions where the number densities are in the range mentioned just above.

5.6 METEOROIDS

Meteoroids are space debris. They range in size from roughly ten microns to about one hundred metres. Larger solid bodies are called asteroids. Meteoroids that are smaller than a few hundred microns are called interplanetary dust particles (IDPs). When meteoroids enter the Earth's atmosphere they create 'shooting stars' or meteors; and if they survive and land on the Earth's surface they are called meteorites.

Meteoric material reaches the Earth, creating spectacular showers and, in rare cases, impressive craters. Meteoric smoke likely has significant effects on some upper mesospheric and lower thermospheric phenomena, including the formation of noctilucent clouds in Polar Regions during the summer. Searches for meteoric material extend to sandy deserts around the world and Antarctica, where the contrast between the albedo of the extra-terrestrial object and that of the sand or snow aids in the discovery of samples. High-altitude flights into the stratosphere have enabled the capture of IDPs and their delivery to Earth.

Meteoric material shows considerable variety. Much of it was processed significantly in the proto-solar nebula. For example, turbulence transported some refractory material into inner regions of the disk where it sublimated and then often transported it back to cooler regions where it condensed. A chondrule is a meteoric inclusion, with a size of a few microns to about a centimetre, which was heated to roughly 2000 K and then cooled within minutes.

5.6.1 Stardust

However, not all of the meteoric material was highly processed and thoroughly mixed together in the proto-solar nebula. Such unprocessed material is identifiable due to the sometimes orders-of-magnitude differences between some of the isotopic abundance ratios in them and the corresponding ratios in interplanetary solids that were fully processed and completely mixed. The pristine material is described as being pre-solar. It contains little information about Solar System chemistry. Rather, pre-solar grains are 'stardust', and the study of them provides insight into the chemistries of stellar outflows. This stardust is not isolated. Instead, pre-solar grains contain between about a billionth and a thousandth of their mass in primitive meteorites and IDPs.

Indications that Solar System solids were not completely mixed were first noticed in the 1950s and early 1960s. Chondrites are stony meteorites that have not been modified by melting or differentiation. Deuterium fractionation in carbonaceous chondrites was found to be atypical and more extreme than could then be explained.

The Path to Planets

Identified pre-solar minerals include diamond, silicates in IDPs, silicates in meteorites, oxides, SiC, graphite and trisilicon tetranitride (Si_3N_4). The preceding list gives the minerals in order of decreasing abundance. The diamond pre-solar grains are the smallest with sizes in the range of 1–25 nm. Many other pre-solar grains range in size from 20 nm to 30 microns.

Meteoriticists and astrophysicists strive to ascertain the type of star in which each type of pre-solar grain formed. To do so, they consider a combination of observations of stellar ejecta and of theoretical understanding of stellar evolution, including nucleosynthesis. For example, graphite pre-solar grains have $^{12}C/^{13}C$ abundance ratios that are larger than the solar ratio. In many of them, the abundances of ^{15}N, ^{18}O and ^{28}Si are enhanced. They also contain the decay products of the radioactive nuclei ^{26}Al, ^{41}Ca, ^{44}Ti and ^{49}V. Consequently, graphite pre-solar grains are believed to have originated in the ejecta of Type II supernovae. Other types of pre-solar grains are thought to have formed in the outflows of red giants and asymptotic giant branch stars.

5.6.2 Amino Acids

The identification of over one hundred different amino acids in chondrites provides a stimulus for speculation concerning the role that the impact of meteorites on Earth may have played in the origin of terrestrial life. Amino acids in chondrites show one to two orders of magnitude more deuterium fractionation than the water found in the same meteorites, a fact which places constraints on explanations of their formation.

One mechanism for forming amino acids involves energy input into an aqueous environment. For example, the radioactive decay of ^{26}Al may heat subsurface water on an asteroid. However, the difference between the deuterium fractionations found in meteoritic amino acids and water rules out such a source of the amino acids.

Experiments have demonstrated that the ultraviolet photolysis of mixtures, like those occurring on interstellar grains, of simple ices, followed by warming, leads to the production of a large variety of amino acids. Prolonged exposure to interstellar ultraviolet radiation may have the same effect on an icy mixture as the shorter exposure to more intense radiation in a laboratory.

Formation of amino acids at temperatures of less than 70 K in interstellar clouds would favour the high level of deuterium fractionation observed in the meteoritic amino acids. However, this picture is not free of problems, the main one being the fact that the interstellar ultraviolet background does not penetrate to the dark regions where interstellar ices are abundant.

Ultraviolet photolysis may have occurred near the surface of the proto-solar nebula if it formed in a region containing stars massive enough to be powerful ultraviolet sources. The proto-solar nebula had a higher density than is typical throughout most of a pre-stellar dense core. Perhaps the higher density would have allowed ice to exist in a sufficiently strong ultraviolet radiation field for the photolysis of ice mixtures to have formed amino acids.

5.7 EXOPLANETS

Two decades after the first full confirmation of the discovery of an exoplanet, searches had led to the identification of many hundreds of planets around stars other than the Sun. Many of these differ markedly from the planets in the Solar System. For example, hot-Jupiters are prevalent. A hot-Jupiter is a giant planet with an orbit that is sufficiently smaller than the orbit of Mercury that the atmospheric temperature of the exoplanet is roughly 1000–2000 K or higher. Some of the hot-Jupiters are amongst the most thoroughly studied exoplanets because their large sizes, compared to Earth's, and small orbits sometimes result in them transiting (*i.e.* passing in front of) their companion stars. During transits, their atmospheres are targets for investigation with transmission spectroscopy.

Of course, a major goal in exoplanet research is the discovery of a planet similar to the Earth. The detection of such a planet is challenging, but searches have revealed super-Earths, which have masses of up to about ten times that of the Earth. Exoplanets with masses between those of a super-Earth and a hot-Jupiter are also known.

The aim in exoplanet exploration that most excites some exoplanetologists is the observation of biomarkers, including particular characteristics of atmospheric chemical composition.

5.7.1 Transmission Spectroscopy of Hot-Jupiters

In 2012 only a handful of the known giant exoplanets having some of the largest measured orbits could be imaged directly. Fortunately, transit spectroscopy makes possible the investigation of hot-Jupiters and some other exoplanets. Due to the high atmospheric temperatures, a large number of levels in molecules like water and methane are populated. The exploitation of this technique in studies of hot-Jupiters has been facilitated by the collaboration of laboratory spectroscopists and researchers using quantum codes to compute the frequencies and transition probabilities of spectral lines.

Transmission spectroscopy has resulted in the discovery of Na, CH_4, CO and CO_2 in hot-Jupiters. The existence of some controversy over the claimed discovery of H_2O reflects the challenges presented by the study of exoplanet atmospheres. The inference of relative abundances of the detected species requires comparisons of data with theoretical models. Transmission spectroscopy has provided evidence of haze, possibly due to the condensation of clouds of refractory species.

5.7.2 Possible Super-Earth Atmospheric Compositions

At least some of the chemistry in an atmosphere with a temperature comparable to those obtaining in the terrestrial atmosphere will not be in thermochemical equilibrium. The CO double bond and N_2 triple bond contribute to the chemical kinetic timescale exceeding the relevant dynamical timescales. Photochemistry in the thin atmospheres of many rocky exoplanets will also drive the chemistry away from such equilibria. Despite the need for kinetic treatments of atmospheric chemistry in detailed models, some interesting speculation is possible.

The Earth's atmosphere is composed primarily of N_2. CO_2 is another important component, though it is less abundant than O_2, which is produced by living organisms. A super-Earth that is several times more massive than the Earth and with an orbit of 1 AU around a star like the Sun might have an atmosphere that differs markedly from that of the Earth. The super-Earth's stronger gravity might prevent outgassed hydrogen from escaping. The most abundant molecule in the super-Earth's atmosphere might be H_2 or H_2O. H_2O and either CH_4 or CO would also be abundant in an H_2-dominated atmosphere.

A hot super-Earth, with an orbit of less than 0.1 AU, might lose such a large fraction of its volatiles that its atmosphere would be composed primarily of silicates, enriched in species such as Ca, Al and Ti.

5.7.3 Potential Bio-markers

The search for life elsewhere in the Universe is, of course, biased by our terrestrial experience. Even here, it is apparent that life can be much stranger than previously thought. "Extremophiles" are now known to exist in regions that are strongly acidic or alkaline, or at very high or very low temperatures, or under great pressure. However, atmospheric O_2 and N_2O are produced by many living organisms and CO_2 is a signature of the terrestrial atmosphere. Hence, certain exoplanet atmospheric molecular compositions would be consistent with the presence of life.

One might suppose that the detection of O_2 in an exoplanetary atmosphere would be hailed as an indication of the presence of life. However, a high O_2 abundance could be due to a runaway greenhouse effect evaporating oceans and subsequent photodissociation of water vapour and the escape of the released hydrogen into space.

Redox chemistry is associated with terrestrial life. This is chemistry in which electrons are gained (reduction) and lost (oxidation). Methane (CH_4) is very reduced, and, unsurprisingly, O_2 is very oxidised. In the Earth's atmosphere, both have abundances that differ by several orders of magnitude from those that would obtain in thermochemical equilibrium. The discovery in an exoplanet atmosphere of a redox pair, such as these two molecules, with abundances far from those occurring in thermochemical equilibrium would, no doubt, be well publicised in the press.

The presence of Earth-like weather would lead to temporal changes in cloud cover. This would cause the albedo of an exoplanet to vary in time, and direct imaging of the planet with weather could reveal the variations. The albedos of parts of the Earth are greatly affected by the presence of life. Vegetation has a particularly high albedo at wavelengths near 750 nm. This facilitates cooling, which prevents the overheating, which would degrade chlorophyll. The unambiguous remote detection of life on exoplanets remains a considerable challenge, but efforts to achieve it continue.

FURTHER READING

P. Caselli, *Planet. Space Sci.*, 2002, **50**, 1133.
J. Cernicharo, M. Guélin, M. Agúndez, K. Kawaguchi, M. McCarthy and P. Thaddeus, *Astron. Astrophys.*, 2008, **478**, L19.
Y. I. Fujii, S. Okuzumi and S. Inutsuka, *Astrophys. J.*, 2011, **743**, 53.
J. Ilee, A. C. Boley, P. Caselli, R. H. Durisen, T. W. Hartquist and J. M. C. Rawlings, *Mon. Not. R. Astron. Soc.*, 2011, **417**, 2950.
K. Lodders and S. Amari, *Chem. Erde*, 2005, **65**, 93.
M. J. Mumma and S. B. Charnley, *Annu. Rev. Astron. Astrophys.*, 2011, **49**, 471.
K. I. Öberg, Q. Chunhua, D. J. Wilner and S. M. Andrews, *Astrophys. J.*, 2011, **743**, 152.
S. Seager, and D. Deming, *Annu. Rev. Astron. Astrophys.*, 2010, **48**, 631.
H. B. Throop, *Icarus*, 2011, **212**, 885.
R. Visser, S. D. Doty, E. F. van Dishoeck, *Astron. Astrophys.*, 2011, **534**, 132.
C. Walsh, T. J. Millar and H. Nomura, *Astrophys. J.*, 2010, **722**, 1607.
C. Walsh, H. Nomura, T. J. Millar and Y. I. Aikawa, *Astrophys. J.*, 2012, **747**, 114

CHAPTER 6
A Universe of Galaxies

6.1 GALAXIES OUTSIDE THE MILKY WAY

There are probably more galaxies outside our Milky Way galaxy than there are stars in the Milky Way. Some of these galaxies are relatively nearby, and we can trace their structures and try to classify them, for example, as 'spirals' or 'ellipticals' or 'irregulars'. We might hope to infer from this classification something about their state of evolution. For these relatively close-by galaxies we might be able to see whether a galaxy has an active nucleus in which material is falling into a central black hole. We might be able to see whether a galaxy is in fact the result of a collision between two galaxies (such as the so-called Antennae galaxies, see Figure **6.1**) where the interstellar content of one may stimulate star formation in the other, a so-called starburst galaxy. But most galaxies are so far away that they are spatially unresolved and we cannot determine their structure (see Figure **6.1** and Table **6.1**).

6.1.1 Types of Galaxies

Galaxies appear in a range of physical states. Some are intrinsically very bright. The brightest may be thousands of times brighter than the Milky Way galaxy which itself has a brightness of around 10^{10} times that of the Sun. Bright galaxies also tend to be the largest. Others may be very faint. Dwarf galaxies may be only a few thousands of times brighter than a modest star such as the Sun.

The Cosmic-Chemical Bond
DA Williams and TW Hartquist
© DA Williams and TW Hartquist 2013
Published by the Royal Society of Chemistry, www.rsc.org

A Universe of Galaxies

Figure 6.1 (a) A giant elliptical galaxy, M 60, and a spiral galaxy, NGC 4647. The spiral galaxy is about the same size as the Milky Way, while the elliptical galaxy is 30% larger and is nearer to the Milky Way. Credit: NASA, ESA, Hubble Heritage Team (STScI/AURA). (b) The irregular galaxy NGC 1427A. Credit: Hubble Heritage Team (AURA/STScI), ESA, NASA. (c) The dwarf galaxy UGC 9128. Credit: ESA/Hubble & NASA. (d) Colliding galaxies: the Antennae. Credit: NASA, ESA, and the Hubble Heritage Team (STScI/AURA) – ESA/Hubble Collaboration. Acknowledgment: B Whitmore (STScI). (e) Hubble Deep Field South. Credit: NASA/ESA.

Table 6.1 Main galaxy classifications, with brief descriptions.

Galaxy type	Description
Ellipticals	Ellipsoidal systems, mildly flattened, supported by the random motions of their stars.
Spirals	Highly flattened disks supported mainly by rotation; gas, dust and stars show a spiral pattern originating at the nucleus.
Barred spirals	Spiral galaxies in which the spiral arms of gas, dust and stars originate at the ends of a bar passing through the nucleus.
Combinations	Most galaxies are a combination of ellipsoids and disks. Where the disk dominates, the ellipsoid is called the 'bulge'; where the ellipsoid dominates, the galaxy is called a 'disky elliptical'. The sequence from pure ellipsoids to pure disks is called the Hubble Sequence, along which the intrinsic galactic properties change systematically.
Dwarfs	Very faint galaxies that do not fall on the Hubble Sequence. Dwarfs without gas and young stars are diffuse, and are called dwarf spheroidals. Dwarfs with abundant gas and active star formation are irregular.
Irregulars	Not dominated by disk or bulge, irregulars lack any obvious symmetry. They may have multiple components and filamentary structure or tails and are associated with mergers or tidal interactions.

Thus, the brightness of galaxies may range over an enormous factor of about a billion.

Galaxies contain interstellar matter. As we have seen in Chapter 4, interstellar clouds are the reservoir of matter that enables star formation to occur. When there is plenty of matter in the reservoir, the star formation may occur at a high rate, and the galaxy must be at an early stage of its evolution. It will contain many young stars and be very bright. But when the reservoir is nearly empty, then the rate of star formation in a galaxy will be very slow, and so as the stars in that galaxy fade and die and are replenished only slowly, the galaxy is reaching the end stage of its evolution. So the fraction of a galaxy's mass that is in the form of gas rather than stars is an important parameter in determining its evolutionary progress. The interstellar medium in the Milky Way galaxy, a spiral galaxy, contains enough mass for the formation of several billion stars of masses comparable to the Sun, and so our galaxy still has the potential for a considerable amount of future star formation. Elliptical galaxies generally have very little interstellar gas; they are also redder in colour, implying that their stars are older and more evolved. Other galaxies may have most of their mass still remaining in the form of gas; they are not so bright, because they have not yet formed large numbers of massive bright stars.

The interstellar medium of a galaxy is therefore important in determining the evolutionary status of that galaxy. Astrochemistry provides molecules that are useful probes of the gas in which they are found. The molecules are also critical in the early stages of star formation since their radiation helps to cool the gas and so promotes gravitational collapse. These molecules are certainly important.

6.1.2 Applying What We Know

Can we simply use what we know about astrochemistry in the interstellar medium of the Milky Way to explore the nature and content of gas in other galaxies? Can we, for example, assume that we can take CO to be a proportional tracer of molecular hydrogen in molecular clouds, as astronomers often do for the Milky Way? Should we expect the chemical richness that we see in the Milky Way to be replicated in all other galaxies? On the other hand, if chemistry in other galaxies is indeed different from that in the Milky Way, in what ways does it differ and how do those differences arise?

The answers to questions like these take us back to Chapter 3, where we discussed the processes that drive the reactions initiating interstellar chemistry. Ionisation by starlight and by cosmic rays, surface and solid state reactions on dust grains, and heating and mixing of gases by gas dynamical processes were seen in Chapter 3 to be important in allowing chemistry to proceed in interstellar clouds and star-forming regions. We have to ask, therefore, do these drivers operate in the same way in all galaxies?

The answer is: clearly not! Some galaxies will be in a phase of rapid star formation and will have many young hot stars. The average interstellar radiation field in such bright galaxies will be many times more intense than in the Milky Way. Other galaxies may have very few of these important hot young stars, and the average radiation field in these dimmer galaxies may be very weak in UV radiation. So the range of intensity in UV radiation may extend over many orders of magnitude, from many thousands of times larger than that in the Milky Way to almost zero in the faint dwarf galaxies.

The acceleration of charged particles to high energies, *i.e.* the origin of cosmic rays, may also be associated with the formation of massive stars. So the ionisation rate of molecular hydrogen induced by cosmic rays in the interstellar media of different galaxies depends on the rate of formation of massive stars in those galaxies. The ionisation rate may also extend over many orders of magnitude.

The chemistry related to dust might also vary from one galaxy to another. Firstly, we cannot be confident of the nature of dust in other galaxies; after all, our knowledge of the formation and nature even of dust in the Milky Way is rather limited. While it is likely that the dust in all galaxies is composed largely of silicates and carbons, the actual nature of those solids and their relative abundance may vary considerably from one galaxy to another. Secondly, the total amount of dust relative to gas may also vary from one galaxy to another, since dust is formed in stars and ejected into interstellar space where it accumulates and is only slowly removed in high-speed shocks. Evolved galaxies may, therefore, have more dust (relative to hydrogen), while galaxies that are at an early stage of evolution may have relatively little dust. So the importance of dust-related chemistry may also vary considerably from one galaxy to another. This affects, in particular, the rate of formation of the hydrogen molecules that are responsible for almost all the chemistry.

The importance of gas dynamical processes in heating and mixing gases may also vary from one galaxy to another. Stellar winds generate those gas dynamics and are capable of heating and mixing gases; they arise in all types of star formation regions, from those involving low mass stars to high mass stars. So while UV and cosmic rays are affected by the formation only of massive stars, dynamically related chemical processes are triggered by the formation of all types of star.

Finally, in addition to the four main drivers of interstellar chemistry, there is another factor that we have not mentioned yet: the relative abundances of the elements. The abundances of the chemically important elements relative to hydrogen in the Milky Way were shown in Table **1.1**. These elements were created in stellar nucleosynthesis and injected into the interstellar medium through stellar winds and explosions (see Chapter **4**), and accumulate there.

The interstellar abundances of these elements relative to hydrogen reflect the nature of the deaths of stars during the lifetime of the Milky Way. But other galaxies may have had longer or shorter periods in which chemically important elements have been accumulated in the interstellar medium, and the types of stars produced may also be different. Therefore, one cannot assume, *a priori*, that the relative abundances of the elements to each other or to hydrogen are necessarily the same as in the Milky Way galaxy.

So, to begin to answer the questions we have posed about the nature of interstellar chemistry in other galaxies we need to have some idea about the sensitivity of the chemistry to changes in the physical parameters such as radiation field and cosmic ray flux. Is there a 'signature set' of molecules associated with galaxies whose chemistry dominated by one (or more) chemical drivers? If so, then we could begin to consider whether it may be possible to determine the essential nature of a galaxy—even if that galaxy is spatially unresolved—simply by looking at the range of molecular species that may be identified in that galaxy. If we could do that, it would be a major achievement for molecular astrophysics.

In the next Section (**6.2**) of this chapter, we will look at the sensitivity of interstellar chemistry to changes in some of the physical parameters from the Milky Way values that we have used in Chapter **4**. Then, in Section **6.3**, we will try to identify molecules that may be abundant in the interstellar medium of one type of galaxy but not in another. Finally, in Section **6.4**, we will take account of the fact that galaxies appear not only as apparently isolated objects but also in immense clusters; for example, the Pisces–Perseus cluster contains over a thousand galaxies. Clusters like Pisces–Perseus can be so densely populated that member galaxies are interacting violently with each other, and the central galaxy accretes matter in the form of gas and stars torn out of galaxies, or even whole galaxies. The conditions in such a cluster may quite be unlike those found in individual galaxies. Is there a distinct molecular signature in galaxy clusters?

6.2 SENSITIVITY OF INTERSTELLAR CHEMISTRY TO PARAMETER VARIATIONS

There are many factors that can affect the chemistry in dense star-forming regions. Observations of a variety of galactic sources seem to suggest that the gas density and temperature in clouds that may form stars are rather similar. So here we shall look at some chemical drivers that may be rather different in different galaxies. These drivers are the ultraviolet radiation field, the flux of ionising cosmic rays, and the abundances of the elements that form molecules relative to hydrogen.

6.2.1 Galaxies with Intense Ultraviolet Radiation Fields

We have seen in Section **3.2.1** that starlight in diffuse clouds can initiate a rather limited chemistry by ionising carbon and some other atoms. Reactions with molecular hydrogen then produce methylidyne (CH) and other radicals, and exchange reactions of these radicals with other atoms lead to the formation of CO and other molecular species. Of course, the radiation field also destroys molecules by photodissociation and photoionisation. So if we increase the intensity of the radiation without increasing the rate of formation of molecules, then we will simply depress the molecular abundances. In particular, if the strong radiation field significantly reduces the abundance of molecular hydrogen, then all the chemistry based on reactions with H_2 will be heavily suppressed.

However, there are situations in the Milky Way as well as in other galaxies where very intense radiation fields impinge on very dense gas: these are the regions where massive stars are forming. Young massive stars generate UV fields that may easily be tens of thousands of times more intense than the mean interstellar radiation field at the point where they impinge on dense gas clouds close to the stars. But if the number density in the gas cloud is high enough, then reactions can occur at high enough rates to offset the rapid loss caused by the intense radiation. This kind of environment is often referred to as a photon dominated region (or PDR; see Section **2.1** and Figure **2.3**).

Does the chemistry in this kind of PDR differ from that in a diffuse cloud (Section **4.1**)? No, there is no major chemical difference, though there is a physical difference, which has related chemical effects. The chemistry in a PDR, as in diffuse clouds, depends on the presence of H_2, and the transition from mainly atomic to mainly molecular hydrogen occurs in diffuse clouds of the Milky Way very near to the edge of the cloud, see Figure **2.8**. But in a PDR with an intense radiation field, this transition occurs further into the cloud, at a position that depends on the radiation intensity (responsible for dissociating H_2), the number density in the gas, and the dust–gas ratio (both of which affect the rate of formation of H_2). For example, the H $\rightarrow$ H_2 transition for the case of diffuse clouds in the Milky Way occurs at an optical depth in the visual of about one tenth, or less. But in a PDR with a UV intensity of tens of thousands times the mean UV intensity in the Milky Way and densities of thousands of times that of diffuse clouds the H $\rightarrow$ H_2 transition occurs at an optical depth of about two. From that depth point, however, reactions of carbon ions with H_2 initiate the chemistry; just as in Milky Way diffuse clouds. So, molecules such as CH, CO, nitrogen monohydride (NH), CN, and even HCN can form.

The major physical differences that occur when the cloud is irradiated by very intense starlight arise because the radiation field is an important energy source for the cloud. Every time that a carbon atom is ionised:

$$C + UV \rightarrow C^+ + e$$

or a grain is ionised in the photoelectric effect:

$$grain + UV \rightarrow grain^+ + e$$

the electron released, e, carries with it energy that is shared in collisions with the gas atoms and molecules, tending to raise the temperature. Thus, PDRs with sufficiently intense radiation fields may have high temperatures. These may reach values as high as 1000 K for intensities of 10^5 times the mean interstellar radiation intensity and a number density of about 10^3 times that of a typical diffuse cloud. At these elevated temperatures, collisions between atoms, ions, and hydrogen molecules are energetic enough to excite oxygen and carbon atoms, carbon ions, and CO molecules to low lying energy levels, from which these species radiate, so cooling the gas and helping to limit the temperature rise. Molecular hydrogen can be both collisionally and radiatively excited, and emits in the near infrared as it jumps down the rotation/vibration ladder of the ground electronic state (see Figures **2.2** and **2.3**). The chemistry in

PDRs is not particularly extensive or sensitive to the radiation intensity, so that from a chemical perspective PDRs are less interesting than some other astronomical objects. But the emissions from H_2 and from low-lying levels of oxygen and carbon atoms and of carbon ions are useful diagnostics of the physical conditions in the PDRs.

The enhanced temperature in PDRs with very strong radiation fields affects the chemistry by opening up reaction channels that, at low temperatures, are inhibited by small barriers. The most important of these channels is the reaction of O with H_2 to form OH, (see Section **3.1.2**). The chemistry in PDRs is strongly depth-dependent, being largely controlled by the transition in carbon-bearing species $C^+ \rightarrow C \rightarrow CO$. At low optical depths, nearly all elements other than hydrogen are in the form of atoms or atomic ions. At higher optical depths in a powerful PDR, the transition to CO as the dominant carbon-bearing form occurs at an optical depth of about five into the cloud, rather than around two in Milky Way molecular clouds. At greater depths in PDRs, the radiation field is considerably weakened by dust absorption, and by optical depths of about ten into the cloud, the chemistry is no longer driven by the radiation field but by cosmic rays.

6.2.2 Galaxies with Intense Fluxes of Cosmic Rays

Starburst galaxies and merging galaxies show extremely high spatial concentrations of young massive stars, at much greater number densities of stars per unit volume than are found in the Milky Way. These regions are believed to be locations where the cosmic ray fluxes are also very much greater than in the Milky Way, possibly as much as ten thousand times greater. As we have seen in Chapter **3**, cosmic rays are very important in driving the chemistry in dark clouds where starlight does not penetrate. So it is tempting to think that simply turning up the knob marked 'cosmic ray flux' will drive chemistry even faster and that some new and interesting chemical phenomena may occur.

But a moment's thought makes us realise that we cannot expect very great chemical complexity to arise this way. For the chemistry in dark clouds is a balance between formation and destruction reactions. Higher cosmic ray fluxes may certainly help the reactions forming any particular species, but they also help the reactions that destroy that same species. For example, while it is certainly true that more cosmic ray ionisation of H_2 leads to more H_3^+, the most important driver of gas phase reactions in dark clouds, more ionisation also means more electrons, so more dissociative recombination occurs:

$$H_3^+ + e \to 3H$$

and the result is that the H_3^+ abundance may change somewhat but does not rise in proportion to the change in the cosmic ray ionisation rate. So, as far as the highly coupled chemical reactions of gas phase chemistry in dark clouds are concerned, there is a complicated interplay, the outcome of which is not immediately obvious.

There is another effect that we must expect in regions of high cosmic ray fluxes. The main heating process in dark clouds arises from the cosmic rays; since starlight does not penetrate, cosmic rays are the only source of energy, other than gas dynamical events occurring close to stars. Each time a cosmic ray ionises a hydrogen molecule, the ejected electron carries away a lot of energy that is shared with the gas in collisions. If there are many more such ionisations than occur in Milky Way clouds, then the rate of heating will rise in proportion, and the temperature will rise from the low Milky Way dark cloud value of around 10 K. Depending on how large it is, this rise may be enough to drive additional reactions that at low temperature are inhibited by barriers. The temperature rise also affects other reactions that have strongly sensitive rate coefficients. So the actual nature of the chemistry in dark clouds may change from that expected in Milky Way clouds. If the temperature increases to such an extent that molecular hydrogen begins to be collisionally dissociated, then all chemistry begins to be suppressed, as we saw in the case of shocks whose velocity exceeds some critical limit (see Section **3.5.1**).

We show in Figure **6.2** the temperature computed in a self-consistent manner with the chemistry (which provides the cooling molecules) for the interior of a dark cloud. The calculation is for a particular cosmic ray ionisation rate which is allowed to take values from about 10^{-17} per second (which is close the value normally assumed for the Milky Way galaxy) up to 10^{-12} per second, some 10^5 times larger. The larger values of the ionisation

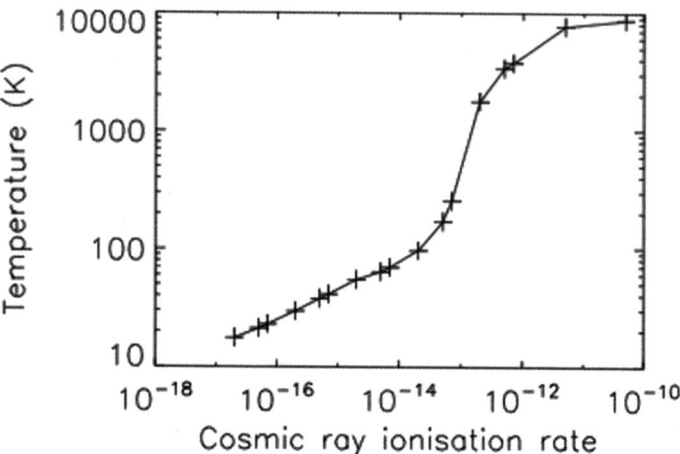

Figure 6.2 Cosmic rays as a heating source inside dark clouds. The cosmic ray ionisation rate is given per second. When the cosmic ray flux is very large, then the temperature of the cloud can reach very high values. Reproduced with permission from E. Bayet *et al.* 2011, Monthly Notices of the Royal Astronomical Society **414** 1583. Copyright: Royal Astronomical Society.

rate may represent regions in starburst and merger galaxies where the number densities of massive stars are very high compared to those in the Milky Way galaxy.

Figure **6.2** shows how important cosmic rays are in controlling the temperature inside dark clouds. For cosmic ray fluxes like those in the Milky Way, the temperatures in dark clouds are low, around 10 K. But an increase of a factor of one hundred or so in the cosmic ray flux raises the temperature to 100 K, at which—for example—all ice accumulation on dust ceases, and all solid state chemistry that generates exceptional chemical complexity is inhibited. If the cosmic ray ionisation rate increases even further, then the temperature can reach about 1000 K. Above that value molecular hydrogen will begin to be dissociated. These dramatic changes in temperature will have important consequences for gas phase chemistry, aside from changes in the abundance of H_3^+, which initiates so much of the chemistry.

We show in Figure **6.3** the chemistry computed consistently with the temperature rise shown in Figure **6.2**.

This figure shows clearly that the abundance of molecular hydrogen is falling dramatically once the cosmic ray ionisation rate

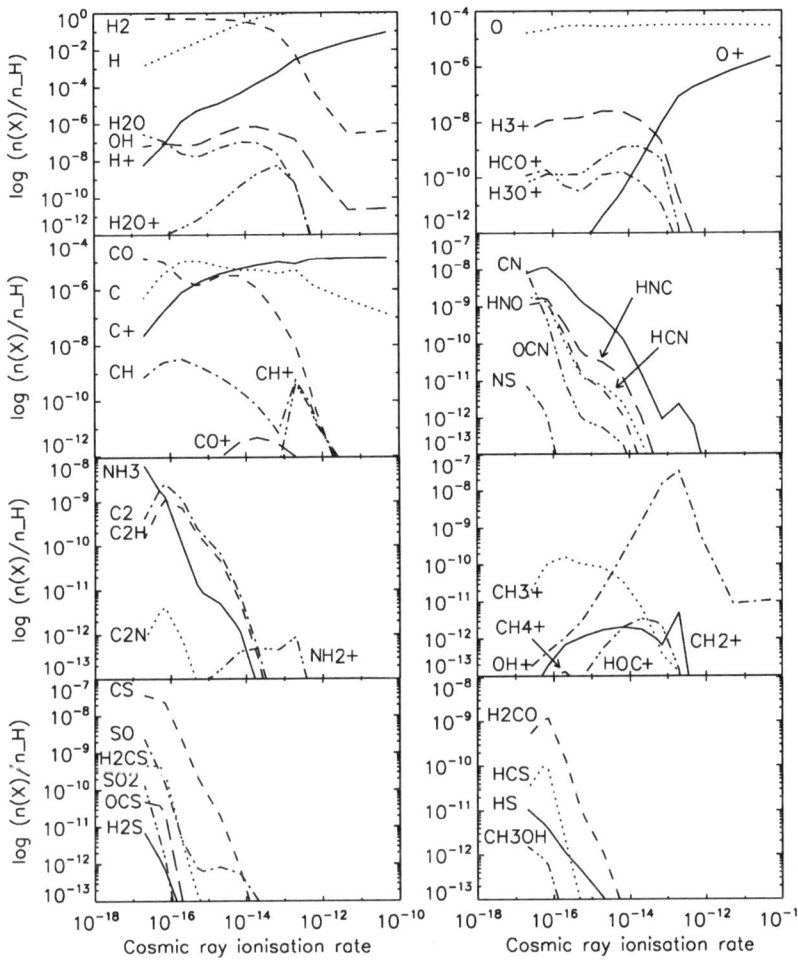

Figure 6.3 The chemistry of a dark cloud depends on the cosmic ray flux. The cosmic ray ionisation rate is given per second. High fluxes of cosmic rays tend to destroy molecules. Reproduced with permission from E. Bayet et al. 2011, Monthly Notices of the Royal Astronomical Society **414** 1583. Copyright: Royal Astronomical Society.

is increased from a value similar to that for the Milky Way, $\sim 10^{-17}$ per second, to about 10^{-14} per second. This makes a consequential reduction in abundance of all neutral species whose chemistry depends on H_2. Nearly all of the neutral species commonly used as tracers of dark cloud gas have abundances that fall steeply as the ionisation rate is increased. The neutral atoms of oxygen and carbon liberated from destroyed molecules are ionised by the cosmic ray flux, and they generate a 'last ditch' chemistry of simple hydride ions with the remaining molecular hydrogen. Ions such as OH^+, CH^+, and CH_2^+ may, therefore, be the best tracers of regions of very high cosmic ray fluxes. Ultimately, however, at the very highest cosmic ray fluxes the elements carbon and oxygen are entirely contained in O^+ and C^+ ions.

6.2.3 Variations in the Relative Elemental Abundances

While star formation is occurring in a galaxy, the mass of interstellar matter is reducing while the total mass of all the stars contained in the galaxy is increasing. But that is not the only change occurring. Massive stars end their lives as supernovae and eject large masses of material into interstellar space (Section **4.4.5**), while novae are similar but smaller explosions (Section **4.4.3**). Cool stars (Section **4.4.2**) and planetary nebulae (Section **4.4.3**) also return material to the interstellar medium. But in all these cases, the material ejected from stars into the interstellar medium is chemically different from the material that formed those objects originally. For the material ejected from supernovae and these other objects contains the 'ashes' of thermonuclear 'burning' in the stellar interiors. In that burning, hydrogen nuclei are converted to more massive atomic nuclei of oxygen, carbon, nitrogen, sulfur, *etc.* Therefore, as the interstellar material is used up in the process of star formation in a galaxy, the remaining interstellar gas becomes enriched in these 'heavy' elements (here, 'heavy' means with atomic masses greater than hydrogen or helium). The abundances of those elements in the interstellar medium relative to hydrogen therefore reflect the evolutionary stage of a particular galaxy. Astronomers use the rather confusing term 'metallicity' to cover the total abundance (relative to hydrogen) of these 'heavy' elements that are capable of forming molecules, *i.e.* all the elements other than hydrogen and the inert gases. A high metallicity means high abundances of oxygen, carbon, nitrogen, sulfur, *etc.* Stellar spectra give a measure of the metallicity of stars of the Milky Way and others have considerably smaller values. Values similar to the Milky Way are common, while some galaxies have metallicities a few times greater and others have considerably smaller values.

How do changes in metallicity affect interstellar chemistry? One might expect that molecular abundances simply scale directly with metallicity. After all, the more oxygen, carbon, nitrogen, *etc.* that is available, the more molecules containing these elements can be made. However, while this is true in an overall sense, it would be a naïve view. We have seen that interstellar chemistry occurs through a large network of reactions forming and destroying molecules. So, an enhancement in, say, the oxygen relative abundance would increase the flux through some reactions (such as the formation of water) and decrease the flux through other reactions (say, the loss of hydrocarbons through oxidation). Some species, therefore, increase with increasing metallicity while other may stay the same or even decrease. Evidently, we cannot use molecular abundances as direct tracers of metallicity in distant galaxies, but if we understand how interstellar chemistry responds to metallicity changes, then we should be able to determine the metallicity in a galaxy from the range of interstellar molecules observed.

We show in Table **6.2** some predictions from computational models of chemistry in photon-dominated regions (PDRs; see Section **2.1**). In these calculations, the PDR is assumed to be irradiated by starlight that is one thousand times as intense as the mean interstellar radiation field in the Milky Way galaxy. The cosmic ray ionisation rate is assumed to be the same as in the Milky Way. The metallicity is varied between the Milky Way value and one percent of that value. The dust–gas ratio and the H_2 formation rate are assumed to scale with the metallicity.

Table 6.2 Sensitivity of molecular fractional abundances to changes in the assumed metallicity (*i.e.* relative abundances of elements other than hydrogen, such as C, N, O, S, *etc.*, capable of forming molecules) in the range from 1–100% of solar. These are computed for a PDR with an intense radiation field (1000× the Milky Way value) but a cosmic ray ionisation rate equal to the Milky Way value. These species are predicted to be detectable.

Molecule	Trends of molecular fractional abundances
CO, H_2O, CS, SO, CO_2, OCS, SO_2, H_2S	Linear tracers of metallicity, Z, for $Z >$ 1% of the solar value.
CN, OH, H_3O^+, HCN, HNC, HCO^+	Fairly insensitive to changes in metallicity.
C_2, C_2H, H_2CS, CH_2CO, H_2CO	Inverse tracers of metallicity, Z, for $Z >$ 1% of the solar value.

Evidently, in PDRs some species (such as CO, CS, SO) are good proportional tracers of metallicity, *i.e.* more metallicity leads to more molecules. Some others (*e.g.* HCN, HNC, HCO$^+$) have abundances that are insensitive to changes in metallicity; while some simple hydrocarbons are—rather surprisingly—inversely dependent on metallicity, *i.e.* more metallicity leads to fewer molecules.

It is interesting from a chemical point of view to understand why this range of behaviours should arise. The reasons are clear when one looks at the chemical network. For example, CO is mainly destroyed in reactions with He$^+$, which is independent of metallicity, but its formation rate increases with metallicity, so overall its abundance increases with metallicity. Reactions forming and destroying OH are both heavily influenced by the abundance of O-atoms, so OH is insensitive to metallicity. Hydrocarbons are lost through a wide variety of exchange reactions, so that a higher metallicity means more loss routes, ensuring their inverse behaviour.

Different results are obtained for physically different regions. For example, in the very dense 'hot cores' (see Section **4.3.1**) close to newly-forming stars, a much richer variety of species arises from the gas-grain chemistry occurring there than from the gas-phase chemistry occurring in PDRs. We show in Table **6.3** similar results to those shown in Table **6.2**, but for hot core chemistry.

Although the range of species in hot cores is different from that in PDRs, the variety of behaviour with respect to changes in metallicity is similar. The abundances of some species (*e.g.* CO, CS,

Table 6.3 Sensitivity of molecular fractional abundances to changes in the assumed metallicity (*i.e.* relative abundances of elements other than hydrogen, such as C, N, O, S, *etc.*, capable of forming molecules) in the range 1–100% of solar. The abundances are computed for a model hot core with hydrogen number density of 10^7 cm^{-3} and a cosmic ray ionisation rate similar to that in the Milky Way.

Molecule	*Trend with change in metallicity*
CS, H$_2$CS	Very strong tracer.
CO, H$_2$O, OCS, SO$_2$, H$_2$CO, CH$_2$CO	Linear tracer.
CH$_3$OH	Insensitive tracer.
SO, HCO$^+$	Inverse tracer.

CH_3OH) follow the metallicity; others (*e.g.* HCN, HNC, CH_3CN) are insensitive to the metallicity, while HCO^+ depends inversely on the metallicity.

6.3 CHEMISTRY IN DIFFERENT GALAXY TYPES

The study of chemistry in galaxies outside the Milky Way is at an early stage, both observationally and in terms of the computational modelling of the chemistry. Nevertheless, it is possible to obtain an indication of how the subject of extragalactic chemistry may develop. Detections of many extragalactic species have already been made, and in the near future we can confidently expect a flood of new molecular detections by telescopes with unprecedented sensitivity, such as ALMA. What can we hope to learn from these observations?

In this section we will consider the chemistry that arises in galaxies in which just one or two drivers dominate the chemistry. We will identify some nearby galaxies whose chemistry may be described in such a way. Then we will describe molecules that are likely to be present in sufficient abundance to be useful tracers of interstellar material.

6.3.1 Galaxies Dominated by Massive Star Formation

We will consider first those galaxies in which the formation of massive stars is occurring at a very high rate. Such galaxies may be at an early phase of evolution in which dense interstellar gas is being rapidly converted to stars. More evolved galaxies may also go through such a phase, known as a 'starburst' phase, if the interstellar medium is abruptly replenished in a collision between galaxies, where one galaxy is rich in interstellar gas. The formation of massive stars is accompanied by the presence of 'hot cores' (see Section **4.3.1**) which will be very rich in molecules, so we expect that galaxies where rapid formation of massive stars is occurring will be powerful emitters in molecular lines from hot core molecules. Even if the galaxies are so far away that individual star forming regions in them cannot be resolved, the radiation from hot core molecules should be detectable.

A relatively nearby galaxy known as M83 (see Figure **6.4**) may be a good example of a starburst galaxy. M83 has a value of the

Figure 6.4 M83: a starbust galaxy. The spiral arms show very many active star-forming regions. (Credit: ESO Science Archive).

metallicity close to that of the Milky Way galaxy. The intense radiation produced by the newly-formed bright stars that it contains does not penetrate the deeply embedded hot core regions because of the huge amount of dust in the cores, and so does not affect the chemistry of the hot core molecules.

Another nearby galaxy also considered to be a starburst galaxy is IC10; it has an irregular morphology. The metallicity of IC10 is only one fifth that of the Milky Way.

The galaxies M83 and IC10 are good candidates to use to test the predictions of chemical models for abundant hot core molecules and the sensitivity of their abundances to the metallicity. Both galaxies are predicted to show very high abundances of many hot core molecules. These include HCN, HNC, CH_3CN, H_2CO, CH_2CO, CH_3OH, CS, H_2CS, H_2S, OCS, SO_2, and SO. The models of these hot cores suggest that larger species such as methyl formate ($HCOOCH_3$) and C_2H_5OH have rather low abundances and it is possible that they may not be detectable in these galaxies. However, the smaller species listed above have remarkably high abundances. In the case of hot cores in M83 and IC10, much of the available sulfur is present in CS, while about one percent of the nitrogen is in HCN and HNC.

In all the cases mentioned here, the abundances for the lower metallicity are less (by one order of magnitude), and sometimes much less (by two orders of magnitude), than for the higher metallicity. In principle, therefore, by observing molecular lines from a family of species (say, N- or S-containing molecules) and calculating their abundances from the intensities one may be able to infer metallicities for distant unresolved galaxies.

6.3.2 Galaxies Dominated by PDRs and XDRs

As we discussed right at the start of this chapter, some galaxies are exceptionally bright in optical and ultraviolet radiation and sometimes in X-rays, too, while others are dim. The intense radiation fields may arise from the rapid formation of massive stars in a starburst galaxy, or from material falling on to a massive black hole at the centre of a galaxy. These latter galaxies are said to have active nuclei. Active galactic nuclei (AGNs) are intense sources of radiation. The very strong gravitational field around one of these nuclei causes material to spiral into the black hole, while frictional forces on the infalling material heat it to very high temperatures, causing it to emit radiation in a wide range of frequencies.

Galaxies with high intensities of UV and X-rays set up PDRs and XDRs. As we have seen (see Section **2.1**), these regions are powerful radiators in molecular lines, in addition to line emissions from any hot core molecules embedded in the star-forming regions.

We discussed the starburst galaxy M83 as a source of molecular line emission from the hot cores it contains. But since it is a starburst galaxy, the young massive stars that it contains generate a very intense ultraviolet radiation field that affects interstellar matter other than that in the very deeply embedded hot core gas. The intense UV radiation creates PDRs in M83 which otherwise has metallicity and other parameters (including the cosmic ray ionisation rate) rather similar to those of the Milky Way. The starburst galaxy NGC253 is similar to M83.

It is interesting to compare the chemistry predicted for these starburst galaxies with predicted results for more quiescent galaxies such as the Milky Way. The more intense radiation fields of the starbursts suppress one group of molecules and enhance another. After CO, the most abundant species in PDRs for Milky Way parameters are CS and H_2O. In starburst galaxies, with their much higher radiation fields, the abundances of these two species are very significantly reduced, but still detectable. Other conventional tracers of interstellar gas such as HCN, HNC, the cyanide radical, and the formyl radical ion (HCO^+) and hydronium ion (H_3O^+) are much less affected by the intense radiation field. Diatomic carbon (C_2) and the ethynyl radical (C_2H) are enhanced in the higher radiation environment, because of the enhanced formation of the carbon ion, C^+.

Galaxies with AGNs are expected to have not only a high intensity of radiation but also a high flux of ionising cosmic rays, compared to the Milky Way. The nearby galaxy NGC3079 appears to have an AGN (see Figure **6.5**).

The additional sources of ionisation in a galaxy with intense radiation fields and enhanced cosmic rays tend to suppress the major sink of carbon, *i.e.* CO and by generating an additional source of carbon ions drive the carbon chemistry. Thus, C_2 and the ethynyl radical are expected to be much more abundant in starburst galaxies with AGNs. Not surprisingly, given the enhanced sources of ionisation in these galaxies, HCO^+ and H_3O^+ are also more enhanced in starburst galaxies with AGNs. Otherwise, most molecular tracers are significantly reduced in abundance.

The observations of molecules in distant galaxies, and the modelling of the chemical networks that produce the molecules, are at an early stage. Nevertheless, it is possible to see that the chemistry responds to the physical parameters associated with different types of galaxy by generating a signature range of molecular species. The corresponding molecular line studies will be capable of identifying the basic properties of individual distant and possibly unresolved galaxies.

6.3.3 Dwarf Galaxies

Very little is known about the molecular content of the interstellar media of dwarf galaxies. CO is at present the only molecule that

Figure 6.5 NGC 3079 is a barred spiral with a very active nucleus. (Credits: NASA, Gerald Cecil (University of North Carolina), Sylvain Veilleux (University of Maryland), Joss Bland-Hawthorn (Anglo-Australian Observatory), and Alex Filippenko (University of California at Berkeley)).

has been detected. The stellar content, however, has been well studied and—rather surprisingly—the stellar populations appear younger than in some large galaxies. This seems to imply that star-formation is on going in some dwarf galaxies and that there may be an ultraviolet component to the interstellar radiation field. If so, then chemistry in dwarf galaxies may be similar to that in larger galaxies.

6.4 CLUSTERS OF GALAXIES

Clusters of galaxies are very large gravitationally bound objects in the Universe. They may contain hundreds of galaxies and have total masses on the order of thousands of billions of solar masses. Between the galaxies in a cluster, there is enough hot gas emitting in X-rays for its mass to be even greater than the total mass of the cluster galaxies themselves. However, the gravity of even this large intracluster mass is not sufficient to hold the cluster of galaxies together, and so astronomers regard this as evidence that 'dark matter' in the cluster provides the necessary gravity to hold the cluster together. The nature of this dark matter remains enigmatic—but it certainly does not make molecules! So we will not consider it further.

Clusters and groups of galaxies are fairly common in the Universe. The Milky Way galaxy itself is a member of a relatively small group of galaxies known as the Local Group which contains about forty galaxies.

The Perseus cluster of galaxies is the brightest object in the X-ray sky and was discovered by a very early X-ray rocket-borne mission in 1970. The Perseus cluster is known to contain nearly two hundred galaxies and is part of the larger Perseus–Pisces cluster, which is about 240 million light years from the Milky Way. The central galaxy in the Perseus cluster is called NGC 1275, and—like the central galaxies of many clusters—is surrounded by huge filaments of gas (see Figure **6.6**). These filaments extend over distances that are comparable with the diameter of the Milky Way Galaxy. The filaments were first detected in optical spectral lines from atomic ions; evidently, the filaments are much cooler than the very hot X-ray emitting gas. However, the origin and the thermal balance of the filaments are unclear at present.

A Universe of Galaxies

Figure 6.6 NGC1275, the central galaxy of the Perseus cluster of galaxies, and its remarkable filaments (Image Credit: NASA, ESA, and L Frattare (STScI).

Later discoveries showed that these filaments contain molecular gas. Molecular hydrogen emission in the infrared traces the optical emission. Millimetre-wave emissions from CO and HCN have also been detected towards the Perseus cluster. High-resolution millimetre-wave observations using arrays of telescopes have shown clearly that the CO is present within some of the filaments associated with NGC1275. These detections indicate that the filaments contain cold molecular gas, and prompted successful millimetre-wave searches for the cyanide radical, HCO^+, and the ethynyl radical.

Once again, the astronomical evidence shows that an extensive chemistry can take place in what appears to be the most challenging and hostile of physical circumstances. It seems remarkable that filaments of cold molecular gas can exist in an environment that is dominated by X-ray emitting gas at a temperature of tens of millions of K. What can we learn about the chemistry of these regions? Is it similar to the chemistry that occurs in molecular clouds in the Milky

Way galaxy? Can we use that understanding to tell us about the physical conditions in the filaments? If so, we may understand better how these filaments arise.

It is possible to make computational models of the filament chemistry. We can imagine that the filament is represented by a slab of gas of which the outer parts that are subjected to UV radiation represent the parts of the filament observed in atomic ions, while the interior, shielded by dust, is the location of the molecules. Of course, the physical parameters of the filament are poorly known, but one can explore the parameter space to see what happens as these parameters (such as gas density, external radiation field, elemental abundances, dust−gas ratio, *etc.*) are varied. The results of extensive calculations using conventional astrochemical networks show that a dominant parameter is the heating rate in the filament. The heating may arise from cosmic rays, which may be intense in such an environment (see Section **6.2.2**) or from a kind of friction occurring as gas flows generated in the intense gravitational fields near NGC 1275 around the filament are shocked (see Section **3.5**).

There is a fairly strong result that comes out of this modelling. If the heating by cosmic rays or gas dynamical friction is not too great, then there is a rich chemistry. The molecules produced in the chemistry radiate in the millimetre waveband and help to cool the gas, so the temperature remains low (typically, less than 100 K). But if the heating is too strong, then the molecules cannot provide enough cooling, so the temperature of the gas rises to such an extent (at least to several thousand K) that any molecules are destroyed. It is almost a flip−flop situation: on the one hand, for heating rates less than a certain limit, there is cool gas and a rich chemistry; on the other hand, above the limit, there is hot gas in which molecules are largely absent. The critical cosmic ray ionisation rate (assuming no frictional heating) is for a value about 10 000 times larger than the cosmic ray ionisation rate in the Milky Way galaxy (see Section **6.2.2**).

Given that the computational models use a chemical network that describes Milky Way galaxy chemistry, it is not surprising that the molecules predicted to arise in the filaments are familiar interstellar molecules. Most of these are simple diatomics (especially CO, CN, and OH) and triatomics (such as HCN, HNC, OCS, OCN, HCO^+), with rather fewer larger species (such as H_2CO, H_2CS, NH_3). Large species, such as those detected in hot cores, are not expected to arise in these filaments unless star formation is also occurring there.

A computational modelling exercise prompted the search for the cyanide radical, HCO^+, and the ethynyl radical in and near the central galaxy of the Perseus cluster, NGC1275. This search was successful.

6.5 CONCLUSIONS

While extragalactic astrochemistry is still a developing topic, it is evident that the chemistry that we can explore in the Milky Way galaxy occurs also in other more distant galaxies. Although the dominant drivers of the chemistry may be different in other galaxies, and many of the physical parameters may also differ from Milky Way values, the chemical network that has been established for the Milky Way seems applicable to other galaxies.

For nearby fully resolved galaxies, then we can see how the chemistry varies from one region to another, just as we can for the Milky Way. Unresolved galaxies, by contrast, emit molecular line signals that encompass a wide range of physical conditions. However, the chemical differences between, for example, the chemistry of a PDR and of a hot core enables the contributions of many types of region to be disentangled. Thus, chemistry provides a powerful way of exploring distant unresolved galaxies.

It seems likely that we shall be able to use astrochemistry to help to classify unresolved galaxies in terms of spirals, ellipticals, irregulars, and dwarfs, since different types of galaxy are expected to show a somewhat different chemistry.

Clusters of galaxies, such as the Perseus cluster, show a complex structure of extensive filaments associated with the central galaxy. The origin of the clusters remains uncertain. However, where molecules are observed in the filaments, it is possible to use chemical models to constrain many of the physical parameters of the cluster.

FURTHER READING

E. Bayet, D. A. Williams, T. W. Hartquist and S. Viti, *Mon. Not. R. Astron. Soc.*, 2011, **414**, 1583.

E. Bayet, S. Viti, D. A. Williams, J. M. C. Rawlings and T. Bell, *Astrophys. J.*, 2009, **696**, 1466.

E. Bayet, S. Viti, D. A. Williams and J. M. C. Rawlings, *Astrophys. J.*, 2008, **676**, 978.

E. Bayet, S. Viti, T. W. Hartquist and D. A. Williams, *Mon. Not. R. Astron. Soc.*, 2011, **417**, 627.

CHAPTER 7
The Early Universe

7.1 COSMIC EVOLUTION BEFORE AND DURING RECOMBINATION

Unless galaxies are within a distance of the order of ten megaparsecs of one another, they move away from one another. The Universe is expanding, and it has been expanding ever since it was so hot and so dense that quarks (components of nucleons such as protons and neutrons) were free and nucleons did not exist. When the Universe was about three minutes old, it had cooled to roughly 10^9 K, and protons and neutrons were already abundant. At that time, through nucleosynthesis, these elementary particles produced deuterium and helium nuclei. When the Universe was about 300 thousand years old, it had cooled enough that most of the matter in it ceased to be ionised, as electrons recombined with positively charged nuclei. Though a small residual fractional ionisation remained, the abundance of electrons dropped sufficiently that the Universe became nearly transparent at visual and longer wavelengths. Observations made now in many directions away from the plane of the Galaxy and at centimetre to sub-millimetre wavelengths, detect photons that last interacted, non-gravitationally, with matter just as the Universe recombined.

7.1.1 The Expansion of the Universe

Though, in 1907, Albert Einstein began working on the theory of general relativity, he first published his general relativistic field equations governing four-dimensional space-time in 1915. By 1917 he had applied them to the description of the structure and evolution of the Universe. When doing so, he made several assumptions including that the properties, when averaged over large enough volumes, of the Universe do not depend on position. For the assumptions that he made, he was unable to find reasonable static solutions of his field equations. Einstein did not believe the Universe to be expanding or contracting. As a consequence, he introduced an additional term into his equations. This term is a constant, called the cosmological constant, and the introduction of the term allowed Einstein to find a solution corresponding to a non-expanding and non-contracting Universe.

In 1929 Edwin Hubble showed that all but the nearest galaxies are receding from the Milky Way. If an object moves away from an observer at a speed much less than the speed of light, the wavelength of the light from the object will appear to the observer to be longer, by a fraction equal to the recession speed divided by the speed of light, than it would be if the object were motionless. The term redshift refers to the increase of the wavelength. Hubble used the redshifts of spectral features produced in the galaxies to measure the speeds. The intrinsic luminosity of a Cepheid variable depends on the period of its variability and is the same for all Cepheid variables having equal periods. This allowed Hubble to employ Cepheid variables as so-called 'standard candles' to establish the distances to galaxies. He did this by using the luminosity-period relation, derived from data for relatively nearby Cepheid variables, to infer the intrinsic luminosity of each observed extragalactic Cepheid variable. He also measured the apparent brightness of each observed extragalactic Cepheid variable. From the ratio of the apparent brightness to the intrinsic luminosity he derived the distance. Hubble found a linear dependence between the recession speed and distance. This result was consistent with the Universe expanding in the way predicted with cosmological solutions to Einstein's field equations, excluding a cosmological constant. The result led Einstein to regret his introduction of the cosmological constant.

However, Hubble's finding did not rule out the cosmological constant. It merely showed that its value is not the one required to prevent the Universe from evolving. Since the early 1980s, two major results have pointed to the need for a term or terms, similar

The Early Universe

to that associated with the cosmological constant, in the general relativistic field equations. One is the development of the theory of cosmological inflation, introduced in 1980 by Alan Guth of the Massachusetts Institute of Technology. The other is the discovery that the expansion of the Universe is accelerating, made independently in the late 1990s by two teams. One included Saul Perlmutter of the University of California Berkeley. The other included Adam Riess, who began his important contributions to the discovery while a postgraduate student at Harvard University, and Brian Schmidt of the Australian National University's Mount Stromlo Observatory. Perlmutter, Riess and Schmidt received the 2011 Nobel Prize in Physics for the discovery.

Inflation is a type of expansion that, if it occurred when the Universe was about 10^{-36} to roughly 10^{-34} seconds old, could account for many important properties of the Universe. These include the property that the Universe's mass and energy densities are very close to being critical in the sense that if the Universe were only a bit denser it would eventually cease to expand and would collapse. A process associated with a term similar to the cosmological constant term can drive inflation.

Perlmutter, Riess, Schmidt and their collaborators observed Type Ia supernovae (regarding them as very bright 'standard candles') in very distant galaxies to determine the distances to those galaxies. These observations revealed a remarkable deviation from Hubble's linear expansion law that implies that the recession of distant galaxies is accelerating. The acceleration could be associated with a term similar to the cosmological constant term.

Currently, the Universe is about 14 billion years old, and the average number density of hydrogen nuclei in it is around 0.2 m^{-3}. About 7 billion years ago, two widely separated galaxies were around half as far apart as now and the average hydrogen number density was close to 2 m^{-3}. In 30 billion years, they will be nearly seven times further from one another than now, and the corresponding number density will be about 300 times smaller.

7.1.2 Primordial Nucleosynthesis

Just as sufficient cooling of an initially atomic gas will lead to the formation of molecules, cooling as the Universe expanded led to the end of the era of free quarks and to the formation of protons and neutrons. After the era of free quarks, the Universe continued to expand and cool. Once it had cooled sufficiently that some simple compound nuclei would not dissociate on timescales shorter than the age of the Universe, the protons and neutrons participated in what could be called a 'chemistry of nucleons' to form the first compound nuclei. This chemistry of nucleons in the Early Universe was first investigated in the late 1940s by George Gamow, then at George Washington University, and Ralph Alpher, who was working at Johns Hopkins University. The chemistry of nucleons in the Early Universe can be modelled with rate equations, just like those used to study a system of combustion or atmospheric chemical reactions. In such a rate equation, the rate per unit time at which the number density of a species changes is determined by the difference between formation and removal terms, and a term used to describe the time-dependent change of the number density due to the expansion or contraction caused by the flow of the medium.

Consider an expanding medium that initially contains only the most basic species. More complicated species become abundant, if they are produced on timescales that are shorter than the timescale on which the expansion causes the overall density to drop. However, the complexity of the species is limited if the formation times of increasingly complicated species are longer than the time for the expansion to cause a significant drop in the density. In the three-minute-old Universe, the most basic species were protons and neutrons. The species that formed within the next few minutes on short enough timescales to contain at least about 10^{-14} of the nuclear mass before the expansion caused their production to stop are: ^{2}H or D (deuterium nuclei); ^{3}H (tritium nuclei); ^{3}He; ^{4}He; ^{6}Li; ^{7}Li; ^{7}Be. The hydrogen and helium nuclei are the most abundant of these. ^{4}He contains roughly a quarter as much mass as all hydrogen, and D and ^{3}He each contain several times 10^{-5} as much mass as all hydrogen. Following the primordial nucleosynthesis era, ^{3}H contained around 10^{-7} as much mass as all hydrogen.

Of the three most abundant species, other than ^{1}H, ^{3}He is the most difficult to detect in astronomical sources and ^{4}He has the abundance that is least sensitive to the parameters that characterise a cosmological model. Thus, the deuterium abundance has been used for several decades as a cosmological diagnostic. The primordial ratio of that abundance to the abundance of all

The Early Universe

hydrogen nuclei is inferred from observations of atomic absorption features in the spectra of quasistellar objects at distances that are a large fraction of the size of the visible Universe.

The discovery that the primordial nucleosynthesis products are limited to the simplest of elements presented a significant challenge to astrophysicists. A number of great nuclear physicists and astrophysicists rose to it. During the 1950s the highly imaginative British astrophysicist Fred Hoyle played a leading role in the development of the field of stellar nucleosynthesis. He inspired other important contributors to join the programme to identify how elements such as carbon, nitrogen and oxygen are produced in evolving stars and what processes in supernovae generate the most massive naturally occurring elements, including iron and even uranium. Some results of this impressive enterprise are described in Section **4.4.5**.

7.1.3 Recombination

After the few minutes of primordial nucleosynthesis, the Universe continued to expand and cool. Initially it remained highly ionised. The electromagnetic radiation scattered on the free electrons. Such scattering is called Compton scattering, and it transferred energy between the radiation field and the electrons. The scattering of electrons with ions transferred energy between those species. As long as the mean free path for the Compton scattering remained small enough compared to the product of the speed of light and the age of the Universe, the radiation field, the electrons and the ions remained well coupled. In such cases, a single temperature characterises the velocity distributions of the electrons and ions. That same temperature also characterises the spectrum of the radiation.

The Universe's expansion induced cooling. Eventually, the temperature dropped enough that neutral atoms survived once they were formed by recombination. This occurred when the temperature of the Universe was around 3000 K. However, the Universe did not become totally neutral. Just as rate equations can be used to model nucleosynthesis in the Early Universe, rate equations can be used to calculate how the electron number density and the ion number density evolved as the Universe expanded. The timescale for the removal of electrons by recombination was inversely proportional to the ion number density, which was equal to the electron number density. The ion number density decreased due to recombination and the expansion of the Universe. When the drop in the ion number density led to the timescale for the removal of electrons by recombination to be about the same as the timescale on which the Universe's expansion decreased the number density of ions, recombination slowed. Eventually, it nearly stopped. When it did, the average distance between hydrogen nuclei was of the order of a thousandth of what it is now, and the fractional ionisation was about 10^{-4}. As described later in this chapter, those remaining electrons had a profound influence on the formation of the first stars.

7.1.4 The Microwave Background

When recombination occurred, the mean free path for Compton scattering became greater than the age of the Universe multiplied by the speed of light. Thus, the Universe became transparent at wavelengths less than 91.2 nm, which is the minimum wavelength of ultraviolet light capable of ionising hydrogen atoms. Just before the transparent era began, the radiation field had a black body spectrum corresponding to a temperature of about 3000 K. Here a 'black body' refers to an ideal radiator in which the matter and the radiation are strongly coupled and is in thermal equilibrium. A black body spectrum has a particular shape and depends on the temperature only. Most of the energy of the radiation in a blackbody with a temperature of 3000 K is carried by photons having wavelengths of about a micron.

The Universe becoming transparent did not change the spectrum of the radiation. However, the expansion of the Universe causes the wavelength of light emitted in the past to become longer in proportion to the average distance between hydrogen nuclei. As mentioned in Section **7.1.3**, the average distance between hydrogen nuclei is now about a thousand times greater than it was at recombination and when Compton scattering became unimportant. The photons in the blackbody radiation field now have wavelengths that are about a thousand times longer than they did at recombination. The increase in the wavelengths maintained the black body shape of the spectrum, but it led to a decrease, by a factor of about one thousand, of the temperature characterising the black body spectrum.

In a 1949 theoretical paper on primordial nucleosynthesis, Ralph Alpher and his Johns Hopkins University colleague Robert Herman noted that a black body radiation background with a temperature of 5 K should exist according to the model that they adopted for the Universe's evolution.

In 1965 Arno Penzias and Robert Wilson, who were then working at Bell Laboratories in Holmdel, New Jersey, became the first to make a direct detection of the cosmic microwave background radiation. Their discovery at a wavelength of about 7 cm was serendipitous. However, colleagues at nearby Princeton University had recently begun to build a receiver to use in a search for cosmic background radiation, which they, independently of early researchers, had concluded should exist. Conversations between Penzias and Wilson and the Princeton scientists led to the identification of the radiation detected by the Bell group and to the side-by-side publication of papers by Penzias and Wilson and by the Princeton scientists Robert Dicke, (P.) James (E.) Peebles, Peter Roll and David Wilkinson. The temperature of the blackbody background was found to be about 3 K.

In fact, the cosmic microwave background had been detected indirectly about a quarter of a century earlier, but the relationship of the detection to cosmology had gone either unnoticed or at least uncelebrated. In 1941 Andrew McKellar of the Dominion Astrophysical Observatory in Canada reported the first indirect detection. He had observed optical absorption features of interstellar CN, from which it was possible to derive a rotational excitation temperature of about 2.3 K. If in a two-level system the radiative transitions are much more rapid than collision-induced transitions, the populations of the levels will be characterised by the temperature of the radiation field. In CN the radiative transitions are rapid enough that this is the case. The Nobel Prize winning chemist Gerhard Herzberg of the Canadian National Research Council even noted this CN excitation temperature in his comprehensive book on the spectra of diatomics, which was published in 1950. Despite the fact that the early CN observations did not influence the development of cosmology, studies of CN were ultimately important for the field. The wavelength associated with the CN transition is 2.64 mm, which is much closer than 7 cm is to the wavelength at which the background spectrum now peaks. Observations of CN made in the 1960s through the early 1970s enabled the probing of the background radiation at wavelengths shorter than those that were accessible directly at that time.

The interpretation of background radiation data obtained in the first decade of the third millennium with the space-based Wilkinson Microwave Anisotropy Probe, named after David Wilkinson, has advanced the knowledge of key cosmological parameters tremendously. As this book is being completed in 2012, the space-based Planck observatory is being used to obtain further background radiation data. That data may even allow the investigation of the effect of primordial gravitational waves on the radiation.

7.2 AFTER THE RECOMBINATION ERA

At the time of recombination, the Universe contained inhomogeneities on length scales that were small compared to the speed of light multiplied by the age of the Universe. However, the inhomogeneities had only modest density contrasts with respect to the average density. High-density contrast structures, like galaxies and the stars within them, developed following recombination. The fragmentation of collapsing gas depends on how well it cools. The small fraction of hydrogen that remained ionised after the era of recombination played a key role in the chemistry producing molecules that acted as coolants; these coolant molecules affected the formation of the first stars. The births of the first stars and galaxies provided sources of ultraviolet radiation, which partially reionised the Universe and led to the further production of coolants.

7.2.1 H_2 and HD—First Stars and Galaxies

CO and dust radiate much of the heat generated by the collapse of present-day star forming material. As mentioned in Section **7.1.2**, elements more massive than helium had small fractional abundances in the Early Universe. So the first stars formed in the absence of the coolants that are now important in stellar birth.

Of course, if the collapse of an inhomogeneity drove the temperature of primordial gas back to somewhat above 10^4 K, a small, but non-negligible, fraction of the electrons that remained following recombination would have had sufficiently energetic collisions with the hydrogen atoms to populate excited electronic states. The population of the excited states requires such temperatures because the lowest excited electronic level of atomic hydrogen lies 10.2 eV above the ground level, and 10.2 eV divided by Boltzmann's constant is 1.2×10^5 K. At a temperature of 3×10^4 K, about two percent of the electrons would have enough energy to induce the population of excited electronic states of atomic hydrogen. The radiative decay of these levels produces ultraviolet radiation, which carries energy away from collapsing gas in which it is emitted. So the temperature of star-forming gas would not have risen to above the order of 10^4 K. However, one would expect star formation in gas at such a temperature to differ from that occurring at present in the Milky Way.

The first and second excited rotational levels of H_2 are at energies corresponding to about 170 and 510 K, respectively, above the lowest level. Because it has no dipole moment, even energetic collisions of H_2, in its ground rotational level, with H and He do not lead to the population of the first excited rotational level. However, such collisions do populate the second excited rotational level. If H_2 formed in abundance following recombination, radiation from it would have kept star-forming gas at a temperature of the order of 100 K during most of its collapse. Currently in the Milky Way's interstellar medium, H_2 is produced primarily on the surface of grains (*cf.* Section **2.2.2**). However, in 1961 (M. R.) Coulter MacDowell, of Royal Holloway College, pointed out that H_2 could be produced in dust-free regions by a gas phase sequence of reactions starting with the radiative attachment of electrons to atomic hydrogen.

The sequence of reactions identified by MacDowell is addressed in Section **2.2.1**. In 1968 Jim Peebles and Robert Dicke invoked these reactions in their discussion of the cooling during the formation of globular clusters, which they suggested were the first objects in the Universe to contain stars. Figure **7.1** shows an image of a globular cluster. The Milky Way contains over a hundred globular clusters. Typically such a cluster contains of the order of 10^5 stars, and usually half of them lie within about ten parsecs of

Figure 7.1 Hubble Space Telescope image of the globular cluster NGC 2808. (Credits: NASA, ESA, G. Piotto (University of Padua) and A. Sarajedini (University of Florida)).

the cluster's centre. The Milky Way globular clusters are in the Galaxy's halo. The more intrinsically bright that a star is, the shorter its life will be, and from the observed upper limit to the intrinsic brightness of individual stars in globular clusters, astronomers have inferred that these clusters contain amongst the oldest stellar populations.

One year before Peebles and Dicke considered cooling during global cluster emergence, William Saslaw and David Zipoy, who worked at the University of Cambridge, had suggested that the formation of H_2 after recombination was initiated by the radiative association of H and H^+ (see Section **2.2.1**). The inclusion of the sequences driven by radiative attachment and by radiative association in models of cosmological chemistry shows that before the first stars formed, a fractional abundance of H_2 approaching 10^{-5} was attained throughout most of the volume of the Universe. This was sufficient for H_2 to have acted as an effective coolant limiting the temperature of gas to about 200 K throughout much of the collapse leading to the first stellar birth.

In work begun at the University of Colorado in the early 1980s with Mike Shull and continued with Alex Dalgarno of the Harvard-Smithsonian Center for Astrophysics, Stephen Lepp considered whether HD was an important coolant during the births of the first stars. Unlike H_2, HD has a dipole moment, which means that collisions of it, in its ground rotational state, with H and He can populate its lowest rotation level. Also the fact that D is more massive than H causes the lowest excited rotational level of HD to lie at a lower energy above HD's lowest rotational level than the first excited rotational level of H_2 lies above H_2's lowest rotational level. Consequently, if HD were formed in large enough abundance, its presence in star forming gas would keep the temperature well below 100 K throughout much of the collapse. HD is formed by a charge transfer reaction.

$$D^+ + H_2 \rightarrow HD + H^+$$

The reverse reaction is an important removal mechanism but is endothermic, which leads to fractionation (see Section **4.3.3**). Cosmological chemical studies indicate that before the first stars formed, a fractional abundance of HD of the order of 10^{-9} was attained throughout most of the volume of the Universe.

The incorporation of the chemistry in hydrodynamic models of the collapse of the first star-forming regions is necessary in order to ascertain whether HD cooling was sufficient to affect the thermal evolution and the dynamics. Such computations indicate that it was not. Furthermore, they show that H_2 cooling kept the collapsing gas at a temperature of around 200 K until the number density of hydrogen atoms reached about 10^4 cm^{-3}. At higher number densities, collision-induced de-excitation of the H_2 levels that are important for the cooling is more rapid than the radiative decay of those levels. This results in the level population distribution being in local thermal equilibrium. Then the cooling rate per unit volume due to H_2 scales as the number density of H_2 rather than as the product of the number densities of H_2 and H. The compression-induced heating rate scales more rapidly with number density, and the gas collapsing to become the first stars began to heat as it collapsed further.

7.2.2 Reionisation, H_2 and Star Formation

A key question in astrophysics concerns feedback in star formation. Does the presence of stars stimulate other stars to form or does it prevent stellar birth? If stellar birth is a self-stimulating process, the feedback is said to be positive. Figure **7.2** shows a region where positive feedback seems to have induced 'sequential star formation' in which a 'wave' of star formation propagates through a cloud.

The first stars and galaxies were sources of radiation. Some of the radiation was capable of dissociating H_2, and the destruction of that coolant might have led to a reduction in star formation. However, some of the radiation caused reionisation. The chemical sequences that gave rise to H_2 production in the Early Universe required the presence of charged particles, and in some places the reionisation might have enhanced the production of coolants (*i.e.* H_2) and of stars.

When addressing the issue of feedback in the earliest era of star formation, one must consider the dynamical effects of the ionising radiation. The photoionisation of a region heats it to a temperature of the order of 10^4 K. The heating causes the gas to expand, and consequently material can be effectively evaporated away from where stars would have formed otherwise. This would lead to negative feedback.

Further consideration shows that positive feedback might be a consequence of the expansion of the photoionised gas. The extent of the ionised gas around a radiation source will be limited, because the ionising photons are absorbed as they counteract recombination that occurs in the primarily ionised region. The primarily neutral region surrounding it will be colder. Though the photoionised gas usually expands at a speed that is less than its sound speed, the expansion may be more rapid than the sound speed in the surrounding gas. The expansion will therefore drive a

The Early Universe

Figure 7.2 Hubble Space Telescope image of the star-forming region N90 in the Small Magellanic Cloud. The radiation of the bright blue stars at the centre of the region is ionising and heating the surrounding material. As the heated material evaporates, a shock precedes the ionisation front propagating into the ambient gas. Dark dusty material outside the ionised bubble lies to the upper left and lower right and makes galaxies beyond it appear reddish brown. The youngest stars are still in the process of forming in the dust ridges. Star formation began in the centre of the region but now occurs away from it. (Image credit: NASA, ESA, and the Hubble Heritage Team (STScI/AURA)-ESA/Hubble Collaboration).

shock into the cooler gas. This obviously pushes material around. Feedback will be negative if the shock disperses matter. However, the compression of material by a shock may lead to enhanced molecular formation, cooling and stellar birth.

FURTHER READING

S. C. O. Glover, *Symp. - Int. Astron. Union*, 2011, **280**, 313.
S. Lepp, P. C. Stancil and A. Dalgarno, *J. Phys. B: At., Mol. Opt. Phys.*, 2002, **35**, R57.
M. Petkova and U. Maio, *Mon. Not. R. Astron. Soc.*, 2012, **422**, 3067.
S. Singh, *Big Bang*, Fourth Estate, New York, 2004.
S. Weinberg, *Cosmology*, Oxford University Press, Oxford, 2008.
D. A. Weintraub, *How old is the Universe?*, Princeton University Press, Princeton, 2010.

CHAPTER 8

Why Chemistry is Important for Astronomy

8.1 MOLECULES AS TRACERS

Chemistry makes molecules. The fundamental importance that molecules have had in astronomy is that their emissions indicate the presence of matter that was previously unknown. By tracing previously unsuspected gas, molecules have therefore transformed the view that astronomers have of the Universe. In this section, we discuss several examples of molecules as tracers of astronomical material that was previously unrecognised.

8.1.1 Interstellar Diffuse Clouds

The 1937 detections of molecular line absorptions in diffuse interstellar gas on lines of sight towards nearby bright stars in the Milky Way galaxy were a great surprise to astronomers. Interstellar gas was thought to be too tenuous and too intensely irradiated by ultraviolet radiation from massive stars to be a location in which chemistry could provide detectable abundances of molecules. However, the detection of the methylidyne radical (CH) and the cyanogen radical (CN), and the laboratory identification by A. E. Douglas and Gerhard Herzberg at the University of Saskatchewan in 1941 of CH^+ in the interstellar spectrum showed clearly that this assumption was incorrect, and that there must be regions in the diffuse interstellar gas in which a limited chemistry (at least) can occur (see Section **4.1**).

However, it soon became clear from early theoretical investigations that regions of diffuse interstellar gas with slightly enhanced density, partially shielded from the interstellar radiation field by the embedded dust grains, should favour the formation of molecules. Many versions of interstellar chemistry for diffuse cloud conditions were explored, the most notable early work being that of David Bates (Queen's University Belfast) and Lyman Spitzer Jr. (Princeton University) in 1951. Those authors introduced all the various types of gas phase reactions that are now normally considered to be playing a part in interstellar chemistry, from radiative association to dissociative recombination (see Table **2.1**), and they also speculated about a possible contribution from surface reactions on interstellar dust grains.

However, the chemistry of diffuse clouds could not be properly described until the chemistry of hydrogen, the predominant element, was understood. Working together at the NASA Goddard Space Flight Center, Ted Stecher and David Williams showed in 1966 that molecular hydrogen could be destroyed by starlight. The detailed work of Alex Dalgarno and Arthur Allison at Harvard University and the Smithsonian Astrophysical Observatory implied that this destruction occurred in almost a 'flip-flop' way, so that molecular hydrogen in diffuse interstellar clouds is either low or high in abundance compared to atomic hydrogen. In other words, diffuse neutral interstellar gas is either mostly *atomic* hydrogen—in which regions rather little chemistry occurs—or substantially *molecular* hydrogen (see Figure **2.8**). In those more molecular regions a significant chemistry based on molecular hydrogen could occur. The chemistry based on H_2 that occurs in the diffuse interstellar medium has been described in Chapter **4**.

Interstellar molecular hydrogen was not actually detected until 1970, when George Carruthers at the Naval Research Laboratory in Washington DC obtained data from a rocket-borne UV spectrograph. He found that significant amounts of H_2 were present in the diffuse interstellar medium towards two massive bright stars, γ Velorum and ζ Puppis. Since then there have been many observational studies of interstellar H_2 on many lines of sight and in many different types of interstellar region. The results of these studies are consistent with the view that molecular hydrogen is formed in reactions on the surface of dust grains and destroyed in photodissociation reactions with starlight, as described in Chapter **2**.

Once the issues concerning hydrogen chemistry were understood, then detailed models of interstellar chemistry could be made, and the archetypal study, since emulated by hundreds of researchers, was that made by John Black and Alex Dalgarno at Harvard in 1977. They constructed a detailed model of interstellar gas along the line of sight towards another bright star, ζ Ophiuchi, and were able to match the derived abundances of many atomic and molecular species detected on this line of sight, including lines from rotationally excited molecular hydrogen. However, CH^+ is a pathological case not fitted by their model. High temperature chemistry is required to form CH^+; as we have seen in Section **3.5**, this may be achieved in shocks or turbulence.

Thus, the immense contribution that interstellar molecules have made is to trace the presence of gas previously unsuspected to exist in the interstellar medium. The hydrogen in this gas is substantially molecular, and initiates a chemistry producing some other simple molecular species. The gas is cool and mainly neutral. However, the detection of CH^+ indicates that interstellar gas cannot always be quiescent. From time to time, it must be undergoing shocks or turbulence, so that the effective temperature is raised enough that CH^+ can be formed. Diffuse interstellar gas is, therefore, a lively, dynamic environment.

The simple quiescent view of these diffuse clouds is also challenged by other observations, mostly of atomic ions but also of molecules CH, CH^+, and CN, that show spectral lines that change on timescales as short as a decade or so (see Figure **8.1**). This seems to imply that relatively tiny knots of gas (about 1 AU in size) are present in the diffuse interstellar medium, and these knots move into and out of the line of sight towards the star, causing temporal variations in the line intensity and profile of atoms and molecules in the knot.

We conclude that molecules have traced diffuse interstellar gas of whose existence we would otherwise have been unaware. Molecules have therefore changed astronomers' perception of the Milky Way galaxy.

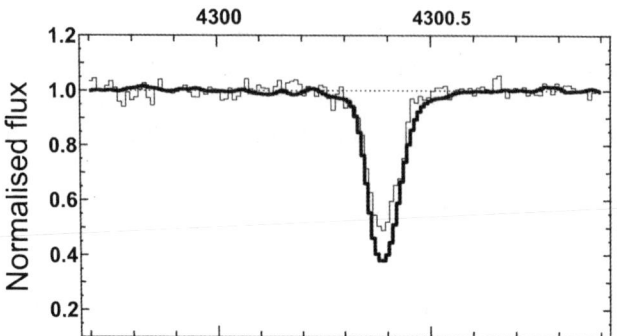

Figure 8.1 Variation in the profile of the 430 nm line of CH along the dusty line of sight towards the star HD 34078, over the period January 1992 (thick line) to February 2002 (thin line). Rollinde *et al.* (2003) show that these data imply that a 20% change in the column density of CH on this line of sight has occurred in less a decade. Reproduced with permission from E. Rollinde *et al.* 2003, Astronomy & Astrophysics **401** 215. Copyright: ESO.

8.1.2 Giant Molecular Clouds

Optical observations of rich star fields show that there are some localised regions where there is an apparent absence of stars. This has been interpreted as the consequence of a concentration of gas and dust that obscures the light of background stars (see, *e.g.* Figure 1.2). It was clear that these rather small regions (known as Bok Globules after the astronomer Bart Bok who catalogued and studied many such objects) should be many times denser than the mean interstellar gas density for the embedded dust grains to provide the necessary optical extinction. But there was no evidence to lead astronomers to believe that dense gas might be extensive throughout interstellar space of the Milky Way or other galaxies.

The detection of interstellar CO in 1970 through its emission in the millimetre-waveband at 2.6 mm revolutionised that picture. CO was found by Arno Penzias and Robert Wilson (who discovered the microwave background, see Section **7.4.1**) and Keith Jefferts (among others) to be present in many objects in the Milky Way galaxy and in external galaxies. By 1977, Phil Solomon, David Sanders and Nick Scoville had identified the so-called Giant Molecular Clouds in the Milky Way, each cloud containing up to a million or so solar masses of material. In the following decade or so Pat Thaddeus, Thomas Dame and collaborators made many extensive CO surveys and completed the first full CO map of the entire Milky Way galaxy at a spatial resolution comparable to that supplied by the atomic hydrogen 21 cm map. It was clear from these studies that the Giant Molecular Clouds in the Milky Way and other galaxies are closely associated with the most important sites of star formation in galaxies.

No molecule has a greater importance in observational astronomy than CO. This molecule is used as a tracer of the much more abundant but usually undetectable molecular hydrogen in many regions of the Milky Way and in dense clouds of gas in external galaxies. CO emission at various wavelengths has been detected in very distant galaxies. In fact, these galaxies may be so distant that they are spatially unresolved and appear simply as a point of light. The current record for the most distant object in which CO emission is detected is from an object so distant that the radiation was emitted when it was at an age of just a few percent of the present age of the Universe. Evidently, CO had already formed in abundance even at this very early epoch in the evolving Universe.

Thus, CO is the astronomical tracer *par excellence*. Through CO, we know where the bulk of matter is distributed in the interstellar space of galaxies. Through CO, the link between interstellar mass and star formation is confirmed. Through CO, the idea is established that interstellar clouds are the reservoirs supplying matter for star formation and therefore controlling the evolution of galaxies.

Emissions from the molecule carbon monoxide established the existence of a huge mass of material whose existence—previously—was completely unknown. Further, without the spur to molecular observations given by using CO as a tracer of mass, molecular astronomy using other molecular emissions would not have been followed up with the same enthusiasm, and the subject of astrochemistry would have been slow to develop.

8.1.3 Interstellar Ice

The idea that ices might exist in the interstellar medium has been much discussed. Indeed, it was widely accepted until the 1960s that interstellar dust grains in the diffuse interstellar medium were made of 'dirty' ice (*i.e.* water ice containing other simple molecules). Bertil Lindblad at Stockholm Observatory pointed out in 1935 that the available elemental abundances favoured the formation of ices. The Leiden astronomers Jan Oort and Henrik van de Hulst in 1946 discussed the nucleation and growth of ice particles and showed that they could provide an excellent fit to extinction measurements in the optical region of the spectrum.

However, these ideas of ice nucleation in the diffuse interstellar medium were incorrect. First, ultraviolet observations of extinction in diffuse interstellar clouds made by Ted Stecher in 1965 indicated that the dust contained solid carbon. When infrared observations became possible, Roger Knacke and collaborators at the University of California at Berkeley showed incontrovertibly that absorption at a wavelength near 3 microns due to the water ice O–H stretch vibration was absent in the diffuse interstellar medium. Apparently, water ice is *not* present in diffuse interstellar gas. The dust in diffuse clouds must be some form of carbon, and also some other material. Silicates were found by Frederick Gillett at Kitt Peak National Observatory in Tucson and collaborators in 1975 to be widespread throughout the entire interstellar medium; they found both absorptions from cold silicates in dense clouds and emissions from hot silicates near stars in the Si–O stretch vibration near 10 microns and O–Si–O bending vibrations near 20 microns. It became accepted, therefore, that dust grains in diffuse clouds were mainly composed of silicates and carbons; this is still the current view.

However, studies on lines of sight through dense clouds—as distinct from diffuse clouds—invariably showed absorption in the ice feature near 3 microns. It appears, therefore, that although ice particles cannot nucleate by themselves in the interstellar medium, interstellar ices can be formed on the surfaces of dust grains made of carbon or of silicates. However, this ice formation occurs only inside dense clouds where the carbon and silicate grains provide significant shielding from starlight. In general, the more shielding there is, the more ice is present, as shown in Figure **3.3**.

The mechanism that controls the onset of ice deposition on dust grains is unclear. This is a bit of surface chemistry that remains to be determined! However, it is clear that the ice must grow not from deposition of H_2O molecules on the surface but from the arrival and hydrogenation *in situ* of oxygen atoms at the surface. Water molecules simply have too low an abundance in the gas phase to allow sufficient ice to accumulate on dust grains in the time available.

Absorption by H_2O ice at a wavelength near 3 microns is an ideal tracer of solid-state material in interstellar clouds. We know now that the ice contains substantial amounts of species other than H_2O (see Section **3.4.2** and Table **3.3**) and that a complex solid-state chemistry can occur in the ice. But H_2O remains the fundamental tracer of solid-state material in the interstellar medium. Where this ice exists, then there is the possibility of greater chemical complexity arising in the relatively simple ices. We conclude that although the early discussions of interstellar ices were incorrect water ice absorption at a wavelength near 3 microns traces precisely those regions where solid-state chemistry may play a role in astronomy.

8.2 MOLECULES AS PROBES

Chemistry makes molecules and—as we have seen in Section **8.1**—molecules trace very precisely the presence of several different types of interstellar material. But molecules are even more valuable to astronomers because the molecular emissions are effective probes of the physical conditions in the gas where they are found. Astronomers can learn about those physical conditions partly from a consideration of the excitation of the various rovibrational levels of the molecules and partly from a detailed computational model of the chemistry that forms the molecules.

8.2.1 Excitation of Molecular Lines

As described in Section **1.4**, there is a limited range of density for which CO molecules in the $J = 1$ rotational level are good emitters. This is true for all molecules in all rovibrational states. If the density is too low, few collisions occur and the intensity of radiation emerging is weak. If the density is too high then collisional de-excitation of the upper rotational level means that the intensity is also weak. But if the collisional rate is comparable to the rate of spontaneous radiation, then the radiation is relatively strong. In the case of CO molecules in a dark cloud at a temperature of 10 K, the number density in the gas for the strongest intensity at a wavelength of 2.6 mm is roughly two thousand hydrogen molecules per cm^3. This density is called the *critical density*. In other words, emission from CO at a wavelength of 2.6 mm is an excellent tracer of gas with density on the order of a thousand H$_2$ molecules per cm^3. But it is a hopeless tracer of cold gas at, say, 100 000 H$_2$ molecules per cm^3, because the high density leads to frequent collisions that de-excite the $J = 1$ level before the molecule can radiate. However, other rotational transitions are useful for tracing gas with different temperatures and densities than those of cold molecular clouds.

For example, the cyanide radical (CN), hydrogen cyanide (HCN), and the formyl radical ion (HCO$^+$) in their low rotational transitions are good tracers of high-density knots of cold gas (say, with number densities around 10^5 H$_2$ molecules per cm^3) in molecular clouds (chemistry in these knots was discussed in Section **4.3**). Similarly, rotational transitions of carbon monosulfide (CS) provide a very useful range of density tracers. In its (2–1) transition CS tends to trace gas at about ten thousand H$_2$ molecules per cm^3, while if we use lines of higher excitation the critical density increases so that in its (7–6) transition CS traces gas at about ten million H$_2$ molecules per cm^3. If we can detect several transitions of the same molecule, then the relative populations in each level of a particular molecule are a good guide to the temperature of the gas: a greater population in higher levels means a higher temperature. The line profiles also tend to become wider at higher temperatures, because the molecules in the gas have a wider distribution of velocities in hotter gas than in cool gas.

Detecting lines from a range of molecules allows astronomers to probe the density and temperature of the region where the molecules are found, as long as we know the critical density of each molecule in the appropriate energy level. Of course, there may be several different components of interstellar gas along the line of sight. The range of molecular tracers and their excitation can help astronomers to unravel these complications. Molecules through their spectral lines provide astronomers with effective probes of interstellar material.

8.2.2 Computational Models of Interstellar Chemistry

Astronomers often want to go further than making estimates of density and temperature of the gas in different regions of interstellar space. For example, it might be important to know the cosmic ray or UV flux in the regions—such as star-forming regions—where the molecules are located (see Section **6.2**); or it might be necessary to determine the elemental abundances in a distant galaxy. The most reliable way of proceeding to determine the physical parameters in a region of space is to make a detailed computational model of the chemistry occurring in that interstellar region.

The chemistry depends on the main drivers (see Chapter **3**), which include cosmic rays, the radiation field, surface chemistry on dust grains, and possibly gas dynamical processes. Any model of chemistry must therefore include the effects of these drivers. To describe the penetration of starlight into an interstellar region we have to describe the optical properties of the dust grains, and their abundance, their size range and their chemical makeup affect these. The model must allow for the possibility that atoms and molecules in the gas phase may stick to the surfaces of dust grains, forming ices, so we need to know the values of those sticking probabilities. Finally, the elemental abundances that are available in the region will affect the chemistry, and may differ from those typical of the Milky Way, so the computational model must allow for variations in abundances to be considered. Evidently, there are potentially a very large number of independent parameters to be included in the model.

The chemistry in the model is expressed in the form of mathematical rate equations that describe the rates of formation and loss of particular species. The chemical network is based on the discussions in Chapter **4**, and typically may include several hundred species of atoms, molecules, and ions linked together in several thousand chemical reactions. The rate coefficients for each of these reactions must be included. Fortunately, the rate coefficients for many of the relevant gas phase reactions have been measured in the laboratory. The situation for surface reactions is less satisfactory. Very few of the many possible surface reactions have been studied in detail.

There are several readily available compilations of chemical reactions for astrochemical models (the UDFA website: www.udfa.net; the Ohio website: www.physics.ohio-state.edu/∼eric/research.html; and the KIDA website: kida.obs.u-bordeaux1.fr). These are frequently updated as rate coefficient data are revised. Using these databases, one can select relevant species, assemble an appropriate reaction network for these species, and construct the mathematical rate equations, including the rate coefficients, to be inserted into the computational model. Since the contribution of surface reactions is less well determined than that from gas phase reactions, it may be necessary to define the surface reactions to be included in the network and the mathematical rate equations.

The modelling exercise then consists of adopting values for each of the parameters listed in Table **8.1**, choosing appropriate initial conditions (it is often assumed that hydrogen is largely molecular while other species are atomic) and running the model to obtain outputs in the form of chemical abundances as a function of time. A typical output is shown in Figure **8.2**.

Table 8.1 Parameter choices to be made before a computational model for astrochemistry is run. The selection of the appropriate chemical network is, of course, a pre-requisite.

Parameter	*Comment*
Gas number density	Is this fixed in space and time? If not, what is the initial distribution and how does it evolve?
Gas temperature	Is this determined self-consistently with the chemistry? Or is it fixed in time and space?
Metallicity	Is this solar? If not, what is its value? Is the region metal-poor or (less likely) metal-rich compared to the Milky Way?
Initial conditions	What is the distribution of the elements among atoms, ions and molecules initially? For example, how is carbon distributed initially among C, C^+, CO? Or hydrogen among H, H_2, H^+, H_3^+?
External UV/optical radiation intensity	Is this the mean interstellar radiation field, or some multiple of it? Or does it have a different wavelength distribution from that of the mean field?
Cosmic ray ionisation rate	Does this have the canonical Milky Way value? Or is it more intense?
Size and shape of the region	These affect the penetration of external radiation into the region, and the emission from molecules out of the region.
Gas–dust ratio	Is this ratio as in the Milky Way? It determines the visual extinction at points within the region.
Dust grain light-scattering properties	These affect the shape of the interstellar extinction curve, and are determined by the dust grains' size and chemical composition.
Gas/grain interactions	These affect a grain's charge, surface chemistry and ice deposition.
Ice processing	What solid-state processing of ices will be adopted?

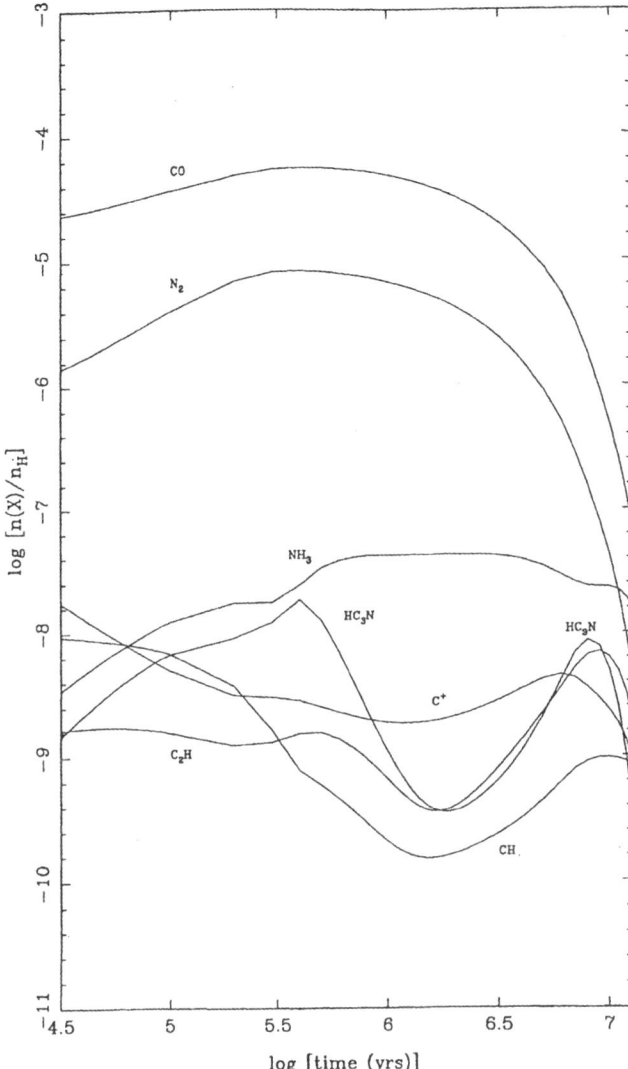

Figure 8.2 A typical output from a computational model of interstellar chemistry. These results show how gas phase chemical abundances of a few species vary with time. The results are computed at a single point deep inside a uniform cloud of total number density of hydrogen 2×10^4 cm^{-3} and temperature 10 K. Initially the gas is assumed to be atomic so that molecules tend to grow in abundance as reactions occur. Ultimately, however, in this model the molecules are frozen on to grain surfaces and removed from the gas, so the abundances decline. Reproduced with permission from D. P. Ruffle *et al.* 1997, Monthly Notices of the Royal Astronomical Society **291** 235. Copyright: Royal Astronomical Society.

For a case in which the gas is initially mainly atomic (apart from H_2 which is assumed to be the main form of hydrogen) the typical output shows a rise in all molecular abundances to peak values depending on the chosen parameters, and then a general decline as the atoms and molecules begin to stick to the surfaces of dust grains. Eventually, if no other processes act to return material from ices to the gas-phase, most species are being lost from the gas and gas-phase chemistry begins to be suppressed.

The aim in the modelling is to vary the input parameters (Table **8.1**) until the predicted abundances in the gas phase and in the ices are comparable to those values implied by the observations. The large number of input parameters suggests that 'solutions' for the physical conditions may not be unique and that the technique is doomed to fail. However, while the number of input parameters listed in Table **8.1** is indeed large, the number of outputs (*i.e.* molecular abundances) is very much larger. The aim in this type of work is to run a variety of models, varying the input parameters listed in Table **8.1** until the outputs, *i.e.* the computed molecular abundances, resemble those obtained from observations. Then we say that these input parameters for that model best represent the physical conditions in the region observed.

8.3 MOLECULES AS CHEMICAL CLOCKS

In earlier chapters of this book we have described the reaction networks that are believed to create the interstellar molecules that have been detected (and also many that have not yet been detected). Each atomic or molecular species in this network is connected via many reactions of various types to other species. If the chemistry that results from such a network is explored computationally as it evolves in time, then—for a specified set of physical conditions—we find a result such as that shown in Figure **8.2**. We see that molecular abundances in the gas grow in time to some value set by the adopted physical conditions, and—in the case described by Figure **8.2**—eventually decline because molecules ultimately stick to grain surfaces and are removed from the gas phase.

It is not surprising that in this evolving situation some molecular species achieve their highest abundance before others. Carbon monosulfide (CS) grows to its maximum abundance at a time well before ammonia (NH_3) reaches its maximum. There are various reasons for this difference in time-dependence. If the important reactions forming a molecular species are fast and involve fairly abundant partners, then the products should grow rapidly in abundance, while routes involving slow reactions with minor partners will give abundances that lag behind the more rapidly formed species. For example, ion–molecule reactions have rate coefficients that are generally large (typically, by a factor of one hundred) compared to those for neutral exchange reactions (and these may also be impeded by activation barriers at low temperatures). Therefore, molecules whose abundances arise from a series of ion-molecule reactions of abundant species rather than neutral exchanges may be expected to arise first in the evolving chemistry.

Of course, there are other considerations that must be taken into account. For example, the gas density is obviously important in determining the speed of the chemistry, because the density determines the rate of collisions between reactants. But for a given density, certain molecular species will appear at an earlier evolutionary time than others. Therefore, molecules can help astronomers to place astronomical objects in an evolutionary sequence. In that sense, molecules can sometimes be used as 'cosmic clock'. Gas containing abundant CS molecules might be at an earlier stage of evolution than gas containing abundant NH_3 molecules. So CS can be called an 'early-time' species, while NH_3 is regarded as a 'late-time' species (as discussed in Section **3.3**). In fact, there are many useful molecular tracers in each category.

The idea that time-dependence in interstellar chemistry is important has allowed astronomers to understand high-resolution studies of molecular clouds. For example, Oscar Morata, Josep Miquel Girart, and Robert Estalella in Barcelona made detailed observational studies of structure in a cloud called L673; this cloud is very clumpy on a rather small (and previously undetected) scale. But, as we have seen in Figure **4.5**, the clumps seen in emission from one molecular tracer (CS, 2–1) do not correspond perfectly to the clumpy structure seen in the emission from another tracer molecule (HCO^+, 1–0). The interpretation that is made is that clouds are certainly clumpy, but that these clumps are transient and at different stages of evolution. The clumps form from low-density gas, contract to form a denser object, and then dissipate.

Sam Falle and Tom Hartquist at the University of Leeds showed that such transient structures might arise during the passage of MHD waves through gas with higher magnetic pressure than thermal pressure. It is a very striking achievement of molecular astrophysics that detections of molecular emissions can help astronomers to describe the large scale gas magnetohydrodynamical properties of the interstellar medium.

8.4 MOLECULES AS CONTROLS

The formation of galaxies and of stars and planets are one-way processes that involve the conversion of gas at very low densities and on a rather large scale into a very dense and relatively small-scale object (see Chapters **4**, **5**, and **6**). For example, the material of a star like the Sun may have originated as a low density gas of perhaps one hundred hydrogen atoms per cm^3 in a cloud extending over a couple of light years. Stars like the Sun have an average density of about one million billion billion (10^{24}) hydrogen atoms per cm^3. The density has been enhanced during this process by a factor of ten thousand billion billion (10^{22}), which—in a spherical compression—means that the radius of the original parcel of gas should reduce by a factor larger than ten million. Gravity is the only force that can operate on such a scale so as to achieve such a compression.

But other forces, principally the internal gas pressure, oppose gravity but magnetic pressure and rotation may also be important.

As gravity begins to compress a cloud, in a mechanical sense it does work on the cloud. This is expressed as heat. (Think of pumping up a bicycle tyre: the pump becomes hot because you are doing work on the air in the pump). But a hotter gas has a higher pressure, which tends to oppose the force of gravity. It is essential, therefore, that gas collapsing under gravity has an efficient cooling mechanism, so that the temperature—and the internal gas pressure—remains low throughout the collapse. Since the gas may initially be very cold, with a temperature near 10 K, the temperature must remain very low during much of the collapse.

We described in Section **8.2.1** how molecules such as CO can be excited by collisions with H_2 molecules from the $J = 0$ to the $J = 1$ rotational states, and that spontaneous emission from that excited state gives rise to the characteristic 2.6 mm emission that is the signature of interstellar CO molecules. This radiation is a loss of energy, *i.e.* a cooling mechanism, for the gas. In other words, the radiation that tells astronomers that interstellar gas is present in a certain location is also the radiation that cools the gas and may—under appropriate circumstances—allow gravitational collapse to continue.

While all molecular emissions from interstellar clouds represent cooling mechanisms, only a few species provide significant amounts of cooling. Besides carbon monoxide (including isotopologues $^{13}C^{16}O$, $^{12}C^{18}O$, *etc.*) other important molecular coolants are OH and H_2O. Atoms such as oxygen that have low-lying energy states also contribute. Dust grains, even at a temperature as low as 10 K, are also effective radiators. But in dust-poor or dust-free regions, molecules are by far the most important coolants.

Therefore, astronomers must regard the ability of molecules to cool collapsing gas as the most important of all the roles of interstellar molecules. Without molecules acting as coolants during the formation of galaxies, stars and planets, it is hard to see how the Universe—including the stars and planets and therefore ourselves—could be in its present state.

8.5 CONCLUSIONS

The discovery of molecules in interstellar space has revolutionized astronomers' view of the Universe. Molecules have revealed the presence of previously unsuspected different forms of material in interstellar space. The study of the emissions and the chemistry that give rise to them provides detailed probes not only of the physical conditions in those regions but also of the evolutionary status of the material. Finally, it is now recognised that molecules are required to provide cooling mechanisms for gas in the processes of collapse to form proto-galaxies or proto-stars. The chemistry that produces interstellar molecules is clearly very important for modern astronomy.

FURTHER READING

D. R. Bates and L. Spitzer, Jr., *Astrophys. J.*, 1951, **113**, 441.
J. H. Black and A. Dalgarno, *Astrophys. J. Suppl.*, 1977, **34**, 405.

T. M. Dame, H. Ungerechts, R. S. Cohen, E. J. de Geus, I. A. Grenier, J. May, D. C. Murphy, L.-A. Nyman and P. Thaddeus, *Astrophys. J.*, 1987, **322**, 706.

S. A. E. G. Falle and T. W. Hartquist, *Mon. Not. R. Astron. Soc.*, 2002, **329**, 195.

T. W. Hartquist and D. A. Williams, *The Chemically Controlled Cosmos*, Cambridge University Press, 1995, (an elementary introduction to the idea of chemical control in astrophysics).

CHAPTER 9

Why Astronomy is Important for Chemistry

9.1 ASTRONOMY AS A STIMULUS TO CHEMISTRY

The flood of new identifications of interstellar and circumstellar molecules in the 1970s and 1980s, mainly by detections of their rotational spectra in the millimetre-wave and submillimetre-wave, showed that these detections could no longer be regarded merely as a curious backwater of science. Interstellar molecules were in fact driving a revolution in astronomy (as we described in Chapter **8**) and it soon became clear that the wholehearted support of laboratory and theoretical chemists would be required to understand and exploit the new information.

The ground-breaking theoretical studies by the Harvard chemist William Klemperer with Philip Solomon (in 1972) and Eric Herbst (in 1973) revealed that much of the chemistry that is required to form the detected species is driven in large part by bimolecular ion-molecule reactions, and that the scale of the chemistry was likely to involve hundreds of molecular species interacting in thousands of reactions. The study of these reactions in the context of astronomy—and the fruitful and enjoyable links between chemists and astronomers—led to the development of the new subject of astrochemistry. As the gas phase chemical networks required for astrochemistry became better understood, it was clear that a great variety of types of reaction were involved, including neutral exchanges and ion-electron recombination (see Chapter 3). By default (at first), gas-grain surface reactions were introduced to account for gaps in model predictions (such as the inability of gas phase reactions to account easily for molecular hydrogen formation), and somewhat later the nature of interstellar ices and their chemical processing became a pressing topic.

The demand for data on ion-molecule, on electron-ion recombination, on neutral exchange reactions, on collisional and radiative excitation processes, on surface reactions, and on ice processing became intense. Laboratory and theoretical scientists responded magnificently to meet the needs posed by the molecular data from astronomical observations.

For the last four decades, therefore, astrochemistry has provided a huge stimulus to chemistry by making these demands for proper scientific understanding of thousands of reactions, their products and their rates. These demands have encouraged the development of new laboratory techniques. Concurrently, the dramatic improvement in computing power that is now widely available has enabled realistic fundamental (*i.e.* quantum mechanical) studies of many reaction types to be performed, with reliable conclusions.

In this chapter we celebrate the stimulus that astronomy has provided to chemistry by highlighting some of the important developments in chemistry that have been encouraged by the insistent demands of astrochemistry. These highlights are, of course, merely a very few of the important examples from a huge range of activity, so these descriptions are illustrative rather than comprehensive. They have been chosen because they enabled important advances in astrochemistry to be made. In Section **9.1** we describe groundbreaking experiments that have enabled astrochemists to contemplate using—with remarkable confidence—gas phase chemical networks of literally thousands of reactions. In Section **9.2** we focus attention on some experiments and theory devoted to the study of surface reactions. Finally, in Section **9.3** we discuss some work on the chemical processing of chemically mixed ices to produce more complex species.

9.2 GAS PHASE REACTIONS IN ASTRONOMY

Major advances in astrochemistry have resulted from the development of ingenious techniques to measure the rate coefficients of gas phase reactions.

9.2.1 Ion−Molecule Reactions: the SIFT Technique

As explained above, astrochemical models are based on networks of (typically) thousands of reactions, many of which are ion−molecule reactions. The models require not only the overall rate coefficient for each reaction (defining the rate of loss of the reactants) but also the proportion of the reaction going into different products. For example, helium ions are known to react with molecular nitrogen to give two product channels.

$$He^+ + N_2 \rightarrow N^+ + N + He$$
$$\rightarrow N_2^+ + He$$

Since astrochemical models must follow the chemistry of all species simultaneously, the branching ratio into these channels is required, along with the overall reaction rate coefficient. Modern experiments show that the overall reaction rate coefficient of the above reaction is 1.6×10^{-9} cm^3 s^{-1}, with the first channel occurring with 60% and the second 40% probability. The branching can be much more complicated than a simple binary split. For example, the reaction of a carbon ion with ethylene, an important 'parent-daughter' reaction in circumstellar envelopes (see Section **4.4.2**), has at least five open product channels at low temperature:

$$C^+ + C_2H_4 \rightarrow C_3H_2^+ + H_2$$
$$\rightarrow C_3H^+ + H_2 + H$$
$$\rightarrow C_2H_3^+ + CH$$
$$\rightarrow H_2C_3H^+ + H$$
$$\rightarrow C_2H_4^+ + C$$

and it is necessary to know the branching ratios (these are experimentally determined to be 22%, 5.5%, 5.5%, 66%, and 1%, respectively, in the order given) as well as the total rate coefficient (1.55×10^{-9} cm^3 s^{-1} in this example).

But how are these branchings determined? Up to the mid-1970s, various techniques were used to determine the overall rate coefficient of a reaction, but the branchings could not normally be obtained. These experiments usually involved measuring the decay of reactant ions in a 'flowing afterglow', a 'stationary afterglow' or a 'drift tube'. From the rate of decay of the reactant, the rate coefficient of the reaction could be obtained. But the growth of the products was not studied in these experiments, and so nothing could be said about the product distributions.

David Smith and Nigel Adams (then both at the University of Birmingham) noted a number of disadvantages in experiments of this type. Unwanted ions were certainly present, and these affected the decay. Concurrent reactions with the primary reaction affected its rate. Metastable species, unwanted electrons and photons might also be present and affect the overall rate of the reaction. In response, they developed a new technique called the 'Selected Ion Flow Tube', or SIFT, to eliminate all these problems and to be able to determine the product branching ratios and product reaction rate coefficients.

The SIFT technique involves using an electrostatic mass 'filter' to select a positive ion beam containing only a single type of ion (selected by mass) from a microwave discharge. This selected ion beam is then injected at a low energy into a low-pressure carrier gas (usually helium), in which the ions thermalise and are carried along with the gas as it moves down a flow tube (about a metre long). Neutral reactant molecules can be injected into the flow tube and the rates of injection are measured. Reactions between the selected ions and injected molecules take place during the flow. At the end of the tube the ions in the flow are sampled through another electrostatic mass 'filter', so that the probability of each product ion arising in the reaction can be determined. This type of experiment overcomes the disadvantages of earlier experiments, as noted by Smith and Adams. A schematic diagram of the original SIFT apparatus is shown in Figure **9.1**.

The SIFT technique in the hands of Nigel Adams and David Smith has been exceptionally successful in measuring very large numbers (possibly as many as a thousand) of rate coefficients and branching ratios of ion–molecule reactions, thereby placing astrochemistry and its models on a reasonably firm footing. The contribution of the SIFT technique and its inventors to the

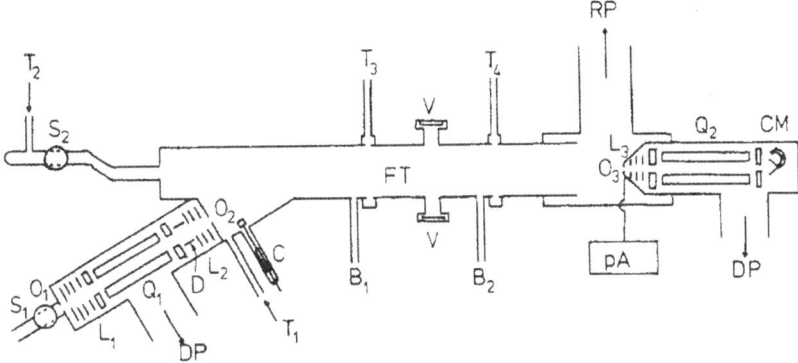

Figure 9.1 A schematic diagram of the SIFT apparatus. S1 and S2 are microwave sources, O1–O3 are orifices, L1–L3 are electrostatic lenses, Q1 and Q2 are quadrupole mass filters, D denotes a deflector plate, C a movable ion collector. FT is the flow tube, T1 and T2 are the carrier gas inlet ports, T3 and T4 are the reactant gas entry ports, B1 and B2 are the pressure measurement ports, V are optical viewing ports, pA is a picoammeter, CM is a channel multiplier, RP is the Roots pump, and DP the diffusion pump. Reproduced with permission from N. G. Adams and D. Smith 1976, Journal of Physics B **9** 1439. Copyright: Institute of Physics.

establishment of astrochemistry as a reliable discipline has been immense.

9.2.2 Neutral Exchange Reactions: the CRÉSU Technique

While many ion–molecule reactions are relatively fast, occurring on nearly every collision, and are often independent of temperature, neutral exchange reactions are generally slower and may have a significant dependence on temperature. The difference occurs because ion–molecule interactions are usually dominated by the strong long-range electrostatic effects between the charge on the ion and the induced (or permanent) dipole in the molecule. On the other hand, much weaker effects such as van der Waals forces may dominate the interactions between neutral species.

Neutral exchange reactions are also likely to be inhibited by any barriers present in the potential energy surface that describes the interaction between the reactants. Any such barriers will be particularly important at the rather low temperatures of interstellar clouds, so that such reactions will be effectively suppressed at very low temperatures. However, reactions between radicals and unsaturated molecules (such as CN + O_2) can be surprisingly fast (this CN, O_2 reaction has a rate coefficient $\sim 10^{-10}$ cm^3 s^{-1}) and their rate coefficients can increase even as temperatures fall to well below 100 K. Reactions between atoms and radicals (such as O + NH) may have no dependence on temperature and have large rate coefficients ($\sim 10^{-10}$ cm^3 s^{-1} in this case). Reactions between saturated molecules are, of course, suppressed at low temperatures because bond breaking is required.

One of the most important techniques for studying the rates of neutral exchange reactions at the low temperatures appropriate for interstellar studies is called CRÉSU (Cinétique de Réaction en Écoulement Supersonique Uniforme, or Reaction Kinetics in Uniform Supersonic Flow). It was devised by Bertrand Rowe and colleagues at the University of Rennes, initially for the study of ion–molecule reactions. It has also been applied with great success to the study of neutral–neutral reactions particularly at very low temperatures by Rowe and by Ian Smith and Ian Sims (originally in the University of Birmingham).

The CRÉSU apparatus relies on a carefully shaped nozzle, a so-called Laval nozzle, to create a flow in the carrier gas that is uniform in temperature, density, and velocity. It is the expansion of the gas in the nozzle that creates a very low temperature in the outflow of gas. In the CRÉSU apparatus, this steady flow can be maintained for several tens of centimetres, during which frequent collisions maintain thermal equilibrium in the gas. Free radicals are created in the flow by pulsed lasers from precursor molecules (for example, H_2O_2 may be used as a precursor for OH) and reactants injected into the flow along with the radical precursors then undergo reactions.

The products of these reactions are observed spectroscopically with a technique called laser induced fluorescence. Lasers are tuned in frequency to excite the product species to high energy levels. The excited products cascade rapidly down to their ground states and radiation from these decays is detected and used as a measure of the abundances of specific products.

The supersonic cooling induced by the expansion of the gas through the Laval nozzle can be extremely effective. Experiments using CRÉSU have measured rate coefficients at temperatures around 10 K. A schematic of the apparatus is shown in Figure 9.2.

The results from the CRÉSU experiments show that neutral exchange reactions at the low temperatures of interstellar clouds

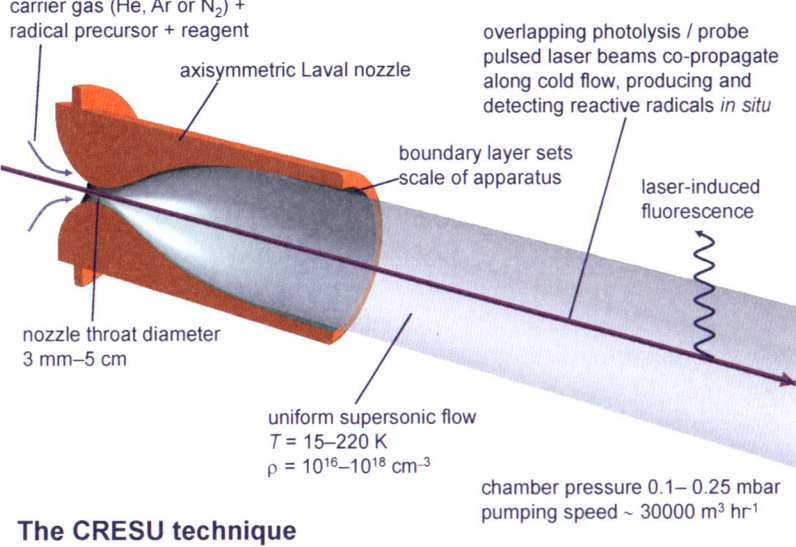

Figure 9.2 A schematic representation of the CRÉSU technique. (Courtesy: I. R. Sims.)

can in some circumstances be exceptionally fast, even rivalling ion-molecule reactions. The developments of this apparatus have made an exceptional contribution to astrochemistry and have shown that the interstellar chemical networks are even more complex than previously assumed.

9.3 SURFACE REACTIONS IN ASTRONOMY

Surface chemists also bring powerful techniques and insight to bear on problems central to astrochemistry.

9.3.1 Molecular Hydrogen Formation

We have described in Chapter 2 how it came to be accepted (initially, by default) that surface reactions on interstellar dust grains were responsible for the formation of interstellar molecular hydrogen. That conclusion was reached in the 1960s, but at that time the idea that one could design an experiment that would actually test such a conclusion in reasonably appropriate physical conditions could not be contemplated because of technical difficulties, and no comprehensive study of this fundamental process could be made. Several decades later, however, technical advances changed that situation dramatically, and several experiments in different laboratories have studied different aspects of surface reactions.

The earliest, most sustained and comprehensive studies of the molecular hydrogen formation process have been made in impressive experiments designed by Gianfranco Vidali (at Syracuse University, New York) and Valerio Pirronello (at the University of Catania, Sicily) and their colleagues, and carried out in Syracuse, N.Y. Their experiments were performed in an ultra-high-vacuum apparatus consisting of a scattering chamber and two beam lines. In their experiments, a beam of H atoms is created by radiofrequency dissociation of molecular hydrogen (H_2) contained in a cavity, and a beam of D atoms is similarly created from the dissociation of deuterium (D_2). These beams impinge and merge on a sample surface in the scattering chamber; this surface represents the interstellar grain surface and is held at a suitably low temperature. The experiment therefore allows the formation of the molecule HD from H and D atoms at the sample; it is assumed that the formation of HD in the experiment mimics the formation of H_2 on interstellar dust, and occurs with similar efficiency. The advantage of using separate H and D beams is that the product molecule (HD) can be distinguished from any background gas (H_2 or D_2) in the apparatus. A mechanical chopper to obtain a lower flux arriving at the sample interrupts the beams. The H atom beam arrives at the sample surface normally, while the D atom beam strikes the sample surface at 38° from the normal. The sample is, if possible, cleaned and baked before being cooled to the desired temperature (typically, on the order of 10 K). The incoming and reflected beams are detected by a 'quadrupole mass spectrometer', an electromagnetic device that 'filters' out particles with a particular mass-to-charge ratio. In these experiments, this detector is set to look for particles of mass 3 atomic units, corresponding to HD. A schematic diagram of the apparatus is shown in Figure **9.3**.

In a typical experiment, the prepared sample (representing the surface of the interstellar grain) is cooled to the desired temperature and then exposed for a pre-determined period to the combined H and D beams. At the end of this period, the sample is quickly warmed to about 30 K, a temperature high enough to desorb any HD on the sample. The detector as a function of time

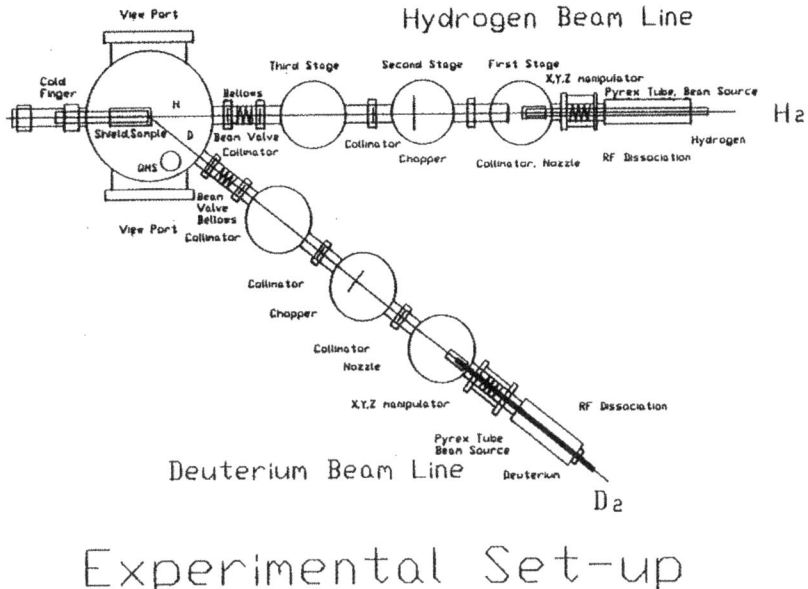

Figure 9.3 A schematic diagram of the H_2 formation apparatus. Reproduced with permission from V. Pirronello *et al.* 1997, Astrophysical Journal (Letters) **475** L69. Copyright: American Astronomical Society.

measures the amount of HD desorbed. From these measurements, and from the incident fluxes of H and D atoms, the overall efficiency of the HD formation process may be determined.

Valerio Pirronello and Gianfranco Vidali, and their collaborators, have measured the formation efficiency on various sample surfaces, including amorphous silicates, amorphous and crystalline carbons, and ices. In general, they find that most of the newly formed HD is retained on the surface until desorption is stimulated by warming of the sample.

Other experimenters have assumed that most newly formed HD will contain considerable internal energy (as rotation and vibration) and will be desorbed from the surface as part of the formation process. Some supporting evidence for that view comes from very detailed theoretical studies based on a quantum mechanical description of the formation of H_2 on a perfect graphite surface. In work done at University College London (UCL) David Clary (now at Oxford University), Anthony Meijer (now at Sheffield

University) and Adam Farebrother computed the energy budget during the formation of molecular hydrogen on a graphite surface. Significant amounts of energy were predicted to appear in internal (rovibrational) modes of the newly formed hydrogen molecules.

These predictions have been confirmed by extensive laboratory studies made by Steve Price and his group at University College London. Molecular hydrogen (HD) formed on a graphitic surface at 15 K is probed by lasers and found in vibrational states from $v'' = 1$ up to $v'' = 7$ and distributed in rotational states up to $J'' = 6$ (however, formation into $v'' = 0$ cannot be measured in the UCL experiment). Much of the available energy released when the H−D bond is formed appears in the surface. The HD translational energy is smaller, but still highly non-thermal for a 15 K surface. These findings suggest that at least some newly formed HD is highly excited internally and is ejected promptly from the surface.

9.3.2 Surface Reactions Forming Species other than H_2

Experiments of this kind have been extensively studied for many years in connection with catalysis, often on metal surfaces. Experiments specifically dedicated to studying surface reactions of interest in astrochemistry have been developed in recent years.

Naoki Watanabe and colleagues at Hokkaido University have performed a very important series of experiments; some of these experiments use the ASURA (Apparatus for Surface Reaction Astrophysics). This consists of an ultrahigh vacuum chamber containing a cold sample (representing the grain surface), an atomic source directed towards the sample, and an infrared spectrometer to interrogate the products of reactions taking place at the surface of the sample.

Naoki Watanabe and Akira Kouchi have studied the formation of CH_3OH and H_2CO by the successive hydrogenation of CO, and have shown that formation of these species is efficient. They also found that reaction intermediates such as HCO, CH_3O and CH_2OH are not detected, consistent with astronomical data. Watanabe and colleagues have made some experiments with ASURA to study the formation of CO_2 and of carbonic acid (H_2CO_3) from OH and CO on an aluminium substrate. Both reactions appear to be efficient.

Another, even more complex, set-up has been developed by Jean Louis Lemaire and colleagues in Paris. This apparatus

(FORMOLISM: FORmation of MOLecules in the InterStellar Medium) includes an ultrahigh vacuum chamber containing a holder for the cold sample representing the dust grain surface, and two atomic/molecular beam lines directed towards the sample. Products can be probed and identified using mass spectrometry, infrared spectroscopy, and by tunable lasers. In an important study using FORMOLISM, it has been demonstrated experimentally for the first time that water can be directly formed on a low temperature substrate representing the surface of an interstellar grain. In their experiment, beams of oxygen and deuterium atoms impinged on an amorphous solid H_2O ice held at 10 K. The efficiency of HDO formation is found to be high.

Evidently, laboratory techniques are now so sophisticated that it is possible to mimic many of the surface reactions that may be occurring in interstellar medium. However, the microscopic physical and chemical nature of the dust grains may be difficult to replicate in the laboratory.

9.4 CHEMICAL PROCESSING OF INTERSTELLAR ICE

Interstellar ices in clouds of moderate density are—chemically— fairly simple. These ices are mainly amorphous solid water, with substantial amounts of carbon monoxide and dioxide, and lesser amounts of formaldehyde, ammonia, methane and other species. It is widely accepted (see Section **4.3.2**) that these molecules are the feedstock for a chemistry that creates the more complex species that are found in regions of star formation. In tiny regions, the so-called hot cores, near to very young stars we find large amounts of relatively complex species such as C_2H_5OH, methyl formate ($HCOOCH_3$), or acetaldehyde (CH_3CHO). Many experiments have been carried out to explore how such a chemical conversion might occur in the laboratory and in space. Generally, an input of energy is required to create radicals or ions that will promote chemistry. In the interstellar medium the likely sources are ultraviolet radiation from a newly formed star or as a result of cosmic ray ionisation of hydrogen (see Section **4.3.2**) or direct ionisation by cosmic rays. For example, in Section **5.6.2** we mentioned that experiments on the UV photolysis of ices produce a variety of amino acids.

Some experimental results obtained by Karin Öberg and colleagues in Leiden were discussed in Chapter **4** (Öberg is now at the University of Virginia). These results were obtained using the CRYOPAD apparatus illustrated schematically in Figure **9.4**.

This is an ultra-high vacuum chamber, operating at $\sim 10^{-10}$ mbar. Ice and ice mixtures are grown up to 100 monolayers thick on a cold substrate, by exposing the substrate to a steady flow of gas. The substrate used is pure gold; this material is inert, and the chemistry is—in any case—dominated by bulk rather than surface properties, so the nature of this substrate should not affect the chemistry. The substrate can be held at any specified temperature between 15 and 200 K. Ultraviolet radiation of a known flux from a hydrogen discharge impinges on the ice mixture for a specified period; the radiation emitted by this discharge ranges from 115–170 nanometres and peaks at 121 nanometres. The resulting ice mixture is then examined by infrared spectroscopy and by a quadrupole mass spectrometer. The apparatus has been used to

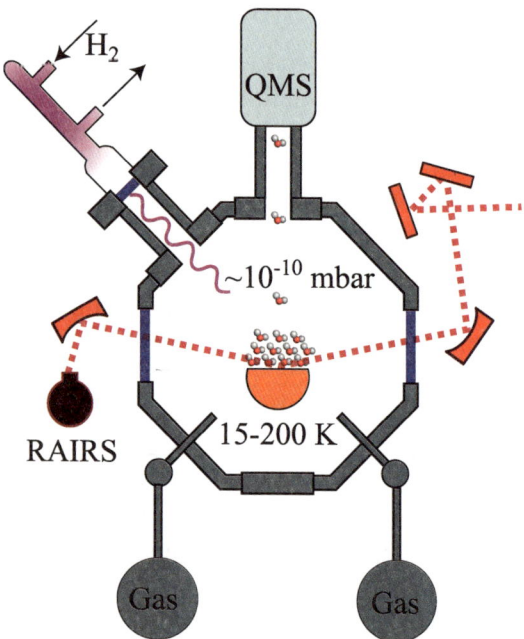

Figure 9.4 A schematic diagram of the CRYOPAD apparatus. Credit: K. Öberg.

determine products of ice processing and their sensitivity to radiation dosage and ice temperatures. It is also used to determine photodesorption probabilities.

9.5 CONCLUSION

Astrochemistry began as a rather speculative subject, based on rather poor astronomical data and with a very limited input of reliable chemical information. The rapid improvement in the astronomical data obtained from sophisticated ground-based and Earth-orbiting telescopes has been paralleled by an astounding development in experimental work in the laboratory. The demands of astronomy for accurate chemical data have been met and continue to be met by remarkable achievements in the laboratory. Astrochemistry is now an experimental science coupled to precise astronomical observations.

FURTHER READING

N. G. Adams and D. Smith, *J. Phys. B: At. Mol. Phys.*, 1976, **9**, 1439.

K. I. Öberg, F. van Broekhuizen, H. J. Fraser, S. E. Bisschop, E. F. van Dishoeck and S. Schlemmer, *Astron. Astrophys.*, 2009, **504** 891; *Astrophys. J, (Lett.)* **621**, L33.

V. Pirronello, O. Biham, L. S. Chi Liu, G. Vidali, *Astrophys. J. (Lett.)* 1997, **483**, L131.

I. R. Sims and I. W. M. Smith, *Annu. Rev. Phys. Chem.*, 1995, **46**, 109.

N. Watanabe and A. Kouchi, *Astrophys. J. (Lett.)*, 2002, **571**, L173.

Subject Index

Note: Page numbers in *italics* refer to figures or tables.

absorption spectroscopy
  Diffuse Interstellar Bands
    (DIBs) 83, *84,* 86–7
  H_2 ultraviolet 22–4
  H_2O ice infrared 64, 190, 191
abundances
  isotope ratios 141
  molecular changes
    during disk
    chemistry 128–9
  in comets 137–9
  H_2O and HCO^+ in evolved
    disks axisymmetric models
    134–6
  and metallicity *159, 160,*
    162–3
  time-dependence of 196–8
  relative elemental 5, 150–1,
    158–61
acetylene 105–6
active galactic nuclei (AGNs)
  163, 164
Adams, Nigel 204–5
AGNs (active galactic nuclei)
  163, 164
A'Hearn, Michael 138
Aikawa, Yuri 133
Allison, Arthur 186

Alpher, Ralph 174, 177
amino acids 4, 137, 141–2
ammonia *see* NH_3
amorphous carbon 33–4
angular momentum 118–20
anions 115, 124
Antennae galaxies 146
aromatic species 107
  see also PAHs (polycyclic
    aromatic hydrocarbons)
associative detachment 29, *46*
astrochemistry 3–4, 201–2
Astronomical Units 19, *21,* 126
ASURA (Apparatus for
  Surface Reaction
  Astrophysics) 210
atmospheres 101, 142–5
axisymmetric models of evolved
  disks 132–6

background microwave
  radiation 176–8
Barnard 68 *7*
barred spiral galaxies *148, 165*
Bates, David 186
benzene 107
biomarkers 144–5
black body radiation 176–7

Subject Index

Black, John 187
Bok Globules 188
Boley, Aaron 130
butterfly-shaped emission regions 73, *74*

C (carbon) 2, 5, 58, 59
 amorphous 33–4
 diamond 141
 ionisation 51, 55, 56
 in novae 109–11
C^+ ions 50–1, 55, 58, 72, 75, 80, 127, 203
C_2 *55*, 111, 138, 164
C_2H (ethynyl radical) 101, 105–6, 164, 167, 169
C_2H_2 101, 138–9
C_2H_4 203
C_2H_6 138–9
Ca, Ca^+ 55
Ca^{++} ions 80
carbon *see* C (carbon); carbon species; carbonaceous dust; hydrocarbon species
carbon dioxide (CO_2) *52*, 64–5, 101, 210
carbon monoxide *see* CO
carbon-rich stars 101, 104–6, *105*
carbon species 81, *105*, 126–7
 see also carbonaceous dust; hydrocarbon species; *and individual species*
carbonaceous dust 7–8, 102–3, 111
carbonic acid (H_2CO_3) 210
carbonyl sulfide (OCS) *52*, 62, 75
Carruthers, George 186
Cepheid variables 172

CH (methylidyne radical) 22, 50–1, 55–6, 81, 187
CH^+ (methylidyne radical ion) 22, 47–9, 72, 158, 187
CH_2^+ 158
CH_3OH (methanol) 66, 75, 96, 129, 210
 in comets 138–9
 proton affinity *52*
CH_4 (methane) *52, 55*, 66, 101, 127
charge transfer reactions 42, 121
chemical clocks, molecules as 61, 196–8
chemical exchange reactions *see* exchange reactions
chemistry drivers 14–15, 149–50
 see also cosmic rays; dust grains; gas dynamical processes; starlight
chondrites *8*, 140, 141–2
chondrules 140
circumstellar regions 2–3, 76–7, 101–3, 104–6
Cl atoms 54–5
Cl^+ ions 56, 80
Clary, David 209–10
clocks, molecules as chemical 61, 196–8
clusters of galaxies 166–9
CN (cyanide/cyanogen radical) 22, 60, 81, 104
 ionisation potential 55
 in Perseus cluster 167, 169
 as tracer 177–8, 192
CO (carbon monoxide)
 conversion to HCO^+ 41, 73–4, 120–1
 in cool star atmospheres 101

as coolant 198–9
in dense cores and disks 126–7
emission spectra *10,* 88, 188–9, 191–2
formation 55, 58, 81, 82, 111, 115
in ice 64–5, 66, 97, *98*
ionisation potential *55*
in Perseus cluster 167
proton affinity *52*
rotational transitions 16–17
sensitivity to metallicity 160
in supernovae ejecta 113, 115
as tracer molecule *10,* 88, 188–9, 191–2
CO_2 (carbon dioxide) *52,* 64–5, 101, 210
collisional dissociation 40, *46*
collisional dissociative ionisation 40, *46*
collisional ionisation 40, *46*
combination galaxies *148*
comets 4, 136–9
complex molecules
in dark clouds 90–2
detected in Milky Way *95*
in diffuse clouds 83–7
in disk formation 129
in planetary nebulae 107
in star-forming regions 66–7, 93–9, 124
see also polyacetylenes; polycyanoacetylenes; polycyclic aromatic hydrocarbons (PAHs)
Compton scattering 175
computational models *see* models and simulations
controls, molecules as 198–9

conversion factors *21*
cool stars and circumstellar envelopes 2–3, 76–7, 101–3, 104–6
cooling mechanisms 17, 178–83, 198–9
cosmic rays 11, 39–41, 51–3, 58–63
in dark clouds 87, 154–8
in diffuse clouds 81–3
variations in different galaxies 150, 154–8, 164
cosmological constant 172–3
cosmological inflation 173
CRÉSU technique 205–7
critical density 192
CRL 618 *107*
CRYOPAD apparatus 212
CS (carbon monosulfide) *52,* 61, 88–9, 104, 164, 192, 196–7
cyanide-bearing species 13, 60–1
see also CN; HCN

Dalgarno, Alex 121, 181, 186, 187
Dame, Thomas 188–9
dark clouds *12,* 58, 87–92, 154–8
effect of shocks 72–3
H_2 in 26–7, 42, 87
ice mantles 64–7, 90–2
dead zones 125
dense cores *11, 12,* 92, 93–9, 126–9
density 9–11, *12,* 20
molecules as probes for 191–2

deuterated species
  H_2D^+ 122–3
  HD 181, 208–9, 210
  HDO (deuterated water) 75, 211
deuterium 2, 174–5
deuterium fractionation 99–100, 122–3
diamond 141
Dicke, Robert 177, 179
diffuse clouds *11, 12,* 78–87, 185–8
  absorption spectra 23–4
  dust surface reactions 53–4, 67–9
  gas phase chemistry 47–53, 55–6, 62–3, 72, 80–3
  H/H_2 balance 42–3
  large molecules 83–7
Diffuse Interstellar Bands (DIBs) 83, *84,* 86–7
disks
  evolved 132–6
  formation of 118–20, 126–9
  massive 129–32
dissociative charge transfer 42, 63
dissociative recombination reactions 38, 41, *46,* 53, 56, 59, 121
DM Tau 125–6
Douglas, A. E. 185
drivers for interstellar chemistry 14–15, 149–50
  see also cosmic rays; dust grains; gas dynamical processes; starlight
dust grains 5–9, 53–4, 150
  formation and composition 7–8, 101–3, 109, 111, 113

  in protoplanetary disks 121–2, 123–4, 125
  size distribution 6–7, 123–4
  surface reactions 53–4
  in diffuse clouds 67–9
  H_2 formation 30–4, 63
  ice mantle formation and processing 64–7
dwarf galaxies 146, *147, 148,* 164–6
dwarf stars, white 14, 108–9, 111

early-time species 61, 197
Early Universe
  expansion 171–3
  primordial nucleosynthesis 173–5
  recombination 175–6
  star formation 178–83
Eddington, Arthur 27
Einstein, Albert 171–2
ejecta
  novae 14, 107–11
  supernovae *57,* 112–16
electromagnetic radiation *see* infrared radiation; starlight; ultraviolet radiation; X-rays
elemental abundance, relative 5, 150–1, 158–61
Eley–Rideal reaction *31,* 32
elliptical galaxies *147,* 148–9
emission spectra 15–17
  CO *10,* 88, 188–9, 191–2
  from comets 137–9
  H_2 24–6, 33, 153–4
  hot silicates 190
  SiO *74*

Unidentified Infrared Bands (UIBs) 83–6
see also tracer molecules
energy, units for 20, *21*
envelopes, circumstellar 2–3, 101–3, 104–6
Estalella, Robert 197
ethynyl radical (C_2H) 101, 105–6, 164, 167, 169
exchange reactions 30, *46,* 49, 197
  ion–molecule exchange 29, 44
  neutral exchange 61, 197, 205–7
exoplanets 142–5

Falle, Sam 198
Farebrother, Adam 210
Fe, Fe^+ *5,* 7–8, *55,* 80
filaments, in Perseus cluster 166–9
formaldehyde (H_2CO) *52,* 66, 130–1, 210
FORMOLISM (FORmation of MOLecules in the InterStellar Medium) 211
formyl radical ion *see* HCO^+
fractional ionisation 120–6
fractionation 99–100, 122–3
FU Orionis 129
fullerenes 107

galaxies
  with active galactic nuclei (AGNs) 163, 164, *165*
  classification 146–9
  clusters 166–9
  dwarf 146, *147, 148,* 164–6
  spiral *10, 147, 148*
  starburst 146, 154, 161–4
  variables affecting chemistry 149–51
    cosmic rays 150, 154–8, 164
    dust 150
    gas dynamical processes 150
    relative elemental abundances 150–1, 158–61, 162–3
    UV radiation 149–50, 152–4, 163–4
  see also Milky Way galaxy; and other individual galaxies
the Galaxy *see* Milky Way galaxy
gamma rays 115, 116
Gamow, George 174
Garrod, Robin 95
gas dynamical processes 69–75, 150
gas phase molecules, in dark clouds 88–90
gas phase reactions and mechanisms
  in diffuse clouds 80–3
  H_2 formation 27–30
  in massive disks 131–2
  rate coefficient measurement 202–7
giant molecular clouds 1, 188–9
Gillett, Frederick 190
Girart, Josep Miquel 197
globular clusters 179–80
grains *see* dust grains
graphite 141
gravitational collapse 92–3, 118–20, 178–82, 198–9
gravitationally unstable disks 129–32

Greenberg, Mayo 97
Guth, Alan 173

H atoms 5, 9
  absence in supernovae ejecta 113–16
  ionisation 35, *37*, *55*, 62
  see also deuterium; tritium
H$^+$ ions 35, *37*, 62, 81
  proton donation 41, *46*, 52
H$_2$ (molecular hydrogen) 22–46
  absence in supernovae ejecta 113–16
  as coolant in Early Universe 179–83
  in dark clouds 26–7, 42, 87, 156–8
  destruction
    by cosmic rays 39–41, 52–3
    by He$^+$ 41–2
    by radiation 35–9, 57–8
  in diffuse clouds 186–7
  exchange reactions 49
  formation
    in gas phase 27–30
    surface reactions 30–4, 63, 207–10
  H/H$_2$ balance 42–3, 87
  importance in interstellar chemistry 44–6
  ionisation 39–40, 52–3, *55*, 57–8, 116
  ion–molecule reactions 41–2, 44, 56, 62
  in novae ejecta 110
  proton affinity 52
  radiative association 49–51
  in shocked regions 71–3
  spectroscopy 22–6, 33, 153–4
H$_2^+$ ions 39–41, 82, 116

H$_2$CO (formaldehyde) *52*, 66, 130–1, 210
H$_2$CO$_3$ (carbonic acid) 210
H$_2$CS (thioformaldehyde) 61, 75
H$_2$D$^+$ 122–3
H$_2$O (water)
  in comets 137–8
  in cool star atmospheres 101
  in dense cores and disks 127, 130–1
  formation 62–3, 71, 82, 211
  as molecular coolant 199
  proton affinity *52*
  in starburst galaxies 164
  see also ices and ice mantles
H$_2$O$^+$ 81
H$_2$S$^+$ 56
H$_3^+$ (protonated molecular hydrogen) 41, 52–3, 58–61, 82, 115–16, 120
H$_3$O$^+$ (hydronium ion, protonated water) 44–5, 53, 81, 164
Hartquist, Tom 198
HC$_3$N (cyanoacetylene) 60
HCl *52*, 56, 81
HCN (hydrogen cyanide) *52*, 60–1, 101, 138–9, 167, 192
HCO$^+$ (formyl radical ion)
  formation 41, 58, 73–5, 82, 120–1
  in massive disks 131, *132*
  in Perseus cluster 167, 169
  in starburst galaxies 164
  as tracer 88–9, 192
HCOOCH$_3$ (methyl formate) 129
HCS (thioformyl radical) 61
HD 181, 208–9, 210

HDO (deuterated water) 75, 211
He, He$^+$ 5, 41–2, 55, 63, 203
heating mechanisms 33, 69–75, 133, 136, 155–8
helium 5, 41–2, 63
Herbst, Eric 95, 121, 201
Herman, Robert 177
Herzberg, Gerhard 177, 185
HNC (hydrogen isocyanide) 60–1, 89
Horsehead Nebula 6
hot atom mechanism 31, 33
hot cores 11, 12, 93, 94, 99, 100, 161, 211
   metallicity tracers in 160–1
hot–Jupiters 142, 143
Hoyle, Fred 175
Hubble, Edwin 172
Hulst, Henrik van de 190
hydrocarbon species 59, 72
   in cool circumstellar envelopes 105
   PAHs 2, 85–7, 102–3, 107, 124
   in planetary nebulae 107
   polyacetylenes 105–6
   sensitivity to metallicity 160
hydrogen see H atoms; H$_2$ (molecular hydrogen)
hydrogen cyanide see HCN
hydrogen isocyanide (HNC) 60–1, 89
hydronium ion (H$_3$O$^+$) 44–5, 53, 81, 164

IC10 162–3
ices and ice mantles
   chemical processing 66–7, 93–9, 211–13
   in dark clouds 64–7, 90–2
   in dense cores 93–9, 126–9
   formation and composition 64–6, 189–91
IDPs (interplanetary dust particles) 139–40, 141
Ilee, John 130
inflation, cosmological 173
infrared radiation
   absorption spectra 64, 190, 191
   by dust 6, 7, 199
   emission spectra 25–6, 137, 138–9, 190
interfaces 73–5
interplanetary dust particles (IDPs) 139–40, 141
interstellar ice see ices and ice mantles
interstellar medium 5–13
   chemistry drivers 14–15, 149–50
   and evolutionary status of galaxy 148–9
   see also dark clouds; diffuse clouds; dust grains
ionisation
   and angular momentum 118–20
   by cosmic rays 39–40, 51–3, 81–3, 150
   by starlight (photoionisation) 36–8, 46, 51, 54–6, 80–1
   by X-rays 57–8
ionisation potentials 55
ion–molecule reactions 29, 44, 46, 197, 203–5
iron species 5, 7–8, 55, 80
irregular galaxies 147, 148

isotopes 2, 141, 174
  see also deuterated species; deuterium fractionation

Jefferts, Keith 188

K, K$^+$ 55
Kepler's supernova *112*
Klemperer, William 121, 201
Knacke, Roger 190
Kouchi, Akira 210
Kuiper Belt 136

L673 88–9, 197
Langmuir–Hinshelwood reaction 30–2
large molecules *see* complex molecules
late-time species 61, 62, 197
Lemaire, Jean Louis 210–11
Lepp, Stephen 181
life, possible evidence for 4, 144–5
Lindblad, Bertil 190
local thermodynamic equilibrium (LTE) 101

M51 9, *10*
M60 *147*
M83 161–3
MacDowell, Coulter 179
magnesium *5, 7,* 80
magnetic fields 17
magneto-rotational instability (MRI) 125
McCoustra, Martin 97
McKellar, Andrew 177
Meijer, Anthony 209–10
merging galaxies 154
metal-containing species *105*

metallic ions 121–2, 123, 124
metallicity 158–61, 162–3
meteoroids 139–42
methane (CH_4) *52, 55,* 66, 101, 127
methanol (CH_3OH) *52,* 66, 75, 96, 129, 138–9, 210
methyl formate ($HCOOCH_3$) 129
methylidyne radical (CH) 22, 50–1, 55–6, 81, 187
methylidyne radical ion (CH^+) 22, 47–9, 72, 158, 187
Mg, Mg$^+$ *5, 7,* 80
microwave background 176–8
Milky Way galaxy 78–117, 148, 166
  complex molecules detected in *95*
  cool circumstellar envelopes 104–6
  dark clouds 87–92
  density 9
  diffuse clouds 78–87
  dust formation 101–3
  novae ejecta 107–11
  planetary nebulae 106–7
  star formation 92–100
  supernovae ejecta 112–16
Millar, Thomas 133
minerals
  pre-solar 141
  silicates 7–8, 141, 190
models and simulations
  axisymmetric disks 132–6
  gravitationally unstable disks 130–2
  interstellar chemistry 192–6
molecular hydrogen *see* H_2
molecules in space

as chemical clocks 61, 196–8
conditions for 14–15
as coolants 198–9
detected species 1–3, 65, 78, 79, 95, 105, 107, 137
formation 47–54
as probes 191–6
role in astronomy 15–17
role in star formation 92–3
as tracers 15–17, 185–91
see also complex molecules
Morata, Oscar 197
MRI (magneto-rotational instability) 125

N atoms 5, 55, 80
N_2 molecules 52, 101, 128, 203
N_2H^+ 89, 129
N90 183
Na, Na^+ 5, 45
near-stellar environments 13–14, 100–16
circumstellar regions 2–3, 76–7, 101–3, 104–6
emission spectra 24–6
novae ejecta 14, 107–11
planetary nebulae 106–7, 108
supernovae ejecta 57, 112–16
neutral exchange reactions 197, 205–7
neutron stars 114
NGC 1275 166–9
NGC 1427A 147
NGC 2403 37
NGC 2808 180
NGC 3079 165
NGC 4647 147
NGC 7538 IRS9 64
NH_3 (ammonia) 68, 89–90, 196–7
in cool star atmospheres 101
in dense cores and disks 128
in diffuse clouds 82–3
in ices 66
in massive disks 130–1
proton affinity 52
nitrogen (N) 5, 55, 80
nitrogen (N_2) 52, 101, 128, 203
nitrogen-bearing species 60–1, 68, 72, 82–3, 105, 106, 107, 128–9
see also cyanide-bearing species; and individual species
NO (nitrogen monoxide) 52, 101, 128
Nomura, Hideko 133
novae ejecta 14, 107–11
nuclear reactions in stars 113–15
nucleosynthesis, primordial 173–5
number densities 9–11, 12, 20

O atoms 2, 5, 8, 55, 62, 80
O^+ ions 44, 62, 81
O_2 molecules 52, 128
Öberg, Karin 96, 212
OCS (carbonyl sulfide) 52, 62, 75
OH 47–9, 62–3, 82, 154, 160, 199
OH^+ 52–3, 158
Oort Cloud 136
Oort, Jan 190
Oppenheimer, Michael 121
optical emission spectra, comets 137–8
oxygen (O) 2, 5, 8, 55, 62, 80
oxygen (O_2) 52, 128

Subject Index

oxygen atomic ion (O^+) 44, 62, 81
oxygen-bearing species 81–2, *105, 107,* 127–8, 141
  see also individual species
oxygen-rich stars 101, 103

P (phosphorus) *105*
PAH hypothesis 85
PAHs (polycyclic aromatic hydrocarbons) 2, 85–7, 102–3, *107,* 124
parent and daughter species 104
parsecs 20, *21*
Peebles, P. James E. 177, 179
Penzias, Arno 177, 188
Perlmutter, Saul 173
Perseus cluster 166, *167*
phosphorus-bearing species *105*
photodetachment 29, *46*
photodissociation 29, 35–9, *46,* 127, 128, 130
photoionisation 36–8, *46,* 51, 54–6, 80–1
photon dominated regions (PDRs) 25, 152–4, 159–60, 163–4
Pirronello, Valerio 208–9
Pisces–Perseus cluster 151, 166
planetary nebulae 106–7, *108*
planets, formation of 118–20
polyacetylenes 105–6
polycyanoacetylenes 13, 60
polycyclic aromatic hydrocarbons (PAHs) 2, 85–7, 102–3, *107,* 124
potassium *55*
pre-solar grains 140–1
Price, Steve 210

primordial nucleosynthesis 173–5
probes, molecules as 191–6
proton affinities 52
proton donation reactions 41, *46,* 52
protonated molecular hydrogen *see* H_3^+
protonated water *see* H_3O^+
protons (H^+) 35, *37,* 62, 81

radiation
  background microwave 176–8
  as cooling mechanism 17, 178–83, 198–9
  see also gamma rays; infrared radiation; ultraviolet radiation; X-rays
radiative association 28–9, *46,* 47–51, 56, 59
radicals 33–4, 97–9, 205–6
  see also individual species
rate coefficients
  measurement of 202–7
  units for 20
RCW 86 57
reaction types *46*
recombination era of Universe 175–6
Red Rectangle *108*
reddening caused by dust 5–6
redshift 172
Riess, Adam 173
Roll, Peter 177
rotational transition spectra 16–17, 26, 191–2
Rowe, Bertrand 206

S, S^+ 54–5, 56, 61, 80

Sanders, David 188
Saslaw, William 180–1
Scoville, Nick 188
Selected Ion Flow Tube (SIFT) 203–5
self-shielding 43
SH (sulfur monohydride) *52,* 56
shielding 43, 109–10
shocks 70–3, 107, 113, 123, 182–3
Shull, Mike 181
Si, Si$^+$ *5,* 7–8, *55,* 80
Si$_3$N$_4$ (trisilicon tetranitride) 141
SiC 141
SiC$_2$ 104
SIFT technique 203–5
silicates 7–8, 141, 190
silicon *5,* 7–8, *55,* 80
silicon-bearing species 7–8, *74,* 101, 104, *105,* 113, 141, 190
Sims, Ian 206
SiO *74,* 101, 113
SiS 104
Smith, David 204–5
Smith, Ian 206
SN1987A 113, 115, 116
SO (sulfur monoxide) 61
SO$_2$ (sulfur dioxide) 61
sodium *5,* 45
solar elemental abundance 5
solar mass 19, *21*
Solar System 126, 136
Solomon, Philip 188, 201
spectroscopy 15–17, 185–92
  comets 137–9
  Diffuse Interstellar Bands (DIBs) 83, *84,* 86–7
  H$_2$ 22–6, 33, 153–4

H$_2$O ice 64, 190, 191
  transmission 143
  Unidentified Infrared Bands (UIBs) 83–6
spiral galaxies *10, 147, 148*
spiral structures in disks 130, 131
Spitzer, Lyman, Jr. 186
star formation 92–100, 189
  in Early Universe 178–83
star-forming regions
  chemical composition compared with comets 139
  complex molecule formation 66–7, 93–9, 124
  deuterium fractionation 99–100, 122–3
  fractional ionisation in 120–6
starburst galaxies 146, 154, 161–4
stardust 140–1
starlight 9–10
  see also photodissociation; photoionisation; ultraviolet radiation; X-rays
stars
  carbon-rich 101, 104–6
  cool 2–3, 76–7, 101–3, 104–6
  neutron 114
  nuclear reactions in 113–15
  oxygen-rich 101, 103
  surface temperatures 3, 13, 101
  white dwarf 14, 108–9, 111
Stecher, Ted 186, 190
sulfur *5,* 54–5, 56, 61, 80
sulfur-bearing species *52,* 61–2, 72, 75, 81, *105, 107*
the Sun 3, 5, 19, *21,* 101
sunspots 3, 101

Subject Index

super-Earths 142, 143–4
supernovae ejecta *57,* 112–16
surface reactions 53–4, 64–7, 67–9, 210–11
  H_2 formation 30–4, 63, 207–10

temperatures
  effect of cosmic rays 155–8
  molecules as probes for 191–2
  stellar surface 3, 13, 101
  see also cooling mechanisms; heating mechanisms
Thaddeus, Pat 188–9
thioformaldehyde (H_2CS) 61, 75
thioformyl radical (HCS) 61
three-body reactions 28, *46,* 76–7, 110
time-dependence of molecular abundances 61, 196–8
tracer molecules *10,* 15–17, 88–90, 185–91
  for metallicity 159–61
  in regions of high cosmic ray fluxes 157–8
  in starburst galaxies 164
transmission spectroscopy 143
trisilicon tetranitride (Si_3N_4) 141
tritium 174

UGC 9128 *147*

ultraviolet radiation 9–10
  in other galaxies 149–50, 152–4, 163–4
  photodestruction of H_2 35–9
  see also photon dominated regions (PDRs)
ultraviolet spectroscopy of H_2 22–6
Unidentified Infrared Bands (UIBs) 83–6
units 19–20, *21,* 126
Universe *see* Early Universe

vibrational energy levels 23, 38
vibrational spectra 64
Vidali, Gianfranco 208–9

Walsh, Catherine 133
warm cores 93
Watanabe, Naoki 210
water *see* H_2O; ices and ice mantles
wavelengths, units for 20, *21*
white dwarf stars 14, 108–9, 111
Widicus Weaver, Susanna 95
Wild 2 4, 137
Wilkinson, David 177
Williams, David 186
Wilson, Robert 177, 188

X-rays 57–8, *112,* 113, 123, 163–4, 166, 167

Zipoy, David 180–1